GIS For Biologists Workbooks
AN INTRODUCTION TO INTEGRATING QGIS AND R FOR SPATIAL ANALYSIS

About The Authors: Dr. Colin D. MacLeod graduated from the University of Glasgow with an honours degree in Zoology in 1994. He then spent a number of years outside of the official academic environment, working as, amongst other things, a professional juggler and magician to fund a research project conducting the first ever study of habitat preferences in a member of the genus *Mesoplodon*, a group of whales about which almost nothing was known at the time. He obtained a masters degree in marine and fisheries science from the University of Aberdeen in 1998 and completed a Ph.D. on the ecology of North Atlantic beaked whales in 2005, using techniques ranging from habitat modelling to stable isotope analysis. Since then he has spent time working as either a teaching or research fellow at the University of Aberdeen and has taught Geographic Information Systems (GIS) at the University of Aberdeen, the University of Bangor (as a guest lecturer) and elsewhere. He has been at the forefront of the use of habitat and species distribution modelling as a tool for studying and conserving cetaceans and other marine organisms and has co-authored over forty scientific papers on subjects as diverse as beaked whales, skuas, bats, lynx, climate change and testes mass allometry, many of which required the use of GIS. In 2011, he created *Pictish Beast Publications* to publish a series of books, such as this one, introducing life scientists to key practical skills and *GIS In Ecology*, to provide training and advice on the use of GIS in marine biology and ecology.

Dr. Ross MacLeod graduated from the University of Glasgow with an honours degree in Zoology in 1999 and went on to study risk trade-off behaviours in birds for his doctoral degree at the Edward Grey Institute of Field Ornithology, University of Oxford, graduating in 2004. Since then, he has focused on behavioural ecology and biodiversity conservation research around the world, and he has led projects in the UK, Bolivia and Peru. After working as a post-doctoral researcher at the University of St Andrews, he moved to the Institute of Biodiversity & Animal Health (IBAHCM), University of Glasgow, to investigate how the impacts of environmental change on biodiversity can be predicted from knowledge of animals behavioural decisions and examining how biodiversity conservation can be delivered through sustainable development and rainforest regeneration. He is currently a lecturer in behavioural ecology at Liverpool John Moores University, focusing on forecasting future population and ecosystem impacts of environmental change. Throughout his career he has been involved in developing new ecological survey techniques and skills-based teaching approaches in the fields of biodiversity measurement, environmental monitoring, statistics and, more recently, GIS.

Cover Image: The oak woodland study area used to study the breeding success of hole-nesting birds, such as redstarts (left) and great tits (right), at the Scottish Centre for Ecology and the Natural Environment (SCENE) field station on the shores of Loch Lomond in Scotland. © Ross Macleod.

PSLS

GIS For Biologists Workbooks
AN INTRODUCTION TO INTEGRATING QGIS AND R FOR SPATIAL ANALYSIS

Colin D. MacLeod and Ross Macleod

Pictish Beast
Publications

Text Copyright © 2019 Colin D. MacLeod and Ross Macleod
Imprint And Layout Copyright © 2019 Colin D. MacLeod/Pictish Beast Publications
PSLS® and THE TOL APPROACH® are registered trademarks

All rights reserved. No part of this book shall be reproduced, stored in a retrieval system, or transmitted by any means, electronic, mechanical, photocopying, recording, or otherwise without permission from the author. No patent liability is assumed with respect to the use of the information contained herein. Although every precaution has been taken in the preparation of this book, the publisher and the author assume no responsibility for errors or omissions. Nor is any liability assumed for damages resulting from the use of the information contained herein.

ISBN – 978-1-909832-52-7
Published by Pictish Beast Publications, Glasgow, UK.
Printed in the United Kingdom.
First Edition: 2019.

Trademarks

All terms mentioned in this book that are known to be trademarks or service marks have been appropriately capitalised. The author and the publisher cannot attest to the accuracy of this information. The use of a term in this book should not be regarded as affecting the validity of any trademark or service mark. In addition, the use of a trademark should not be taken to indicate that the owner of that trademark endorses the contents of this book in any way, or that the author and publisher of this book endorses a particular brand or product.

Warning And Disclaimer

Every effort has been made to make this book as complete and as accurate as possible, but no warranty or fitness is implied. The information provided is on an 'as is' basis and is provided as examples for training purposes only. The authors and the publisher shall have neither liability nor responsibility to any persons or entity with respect to any loss or damages arising from the information contained in this book.

'Space, the final frontier...'
James T. Kirk,
Captain, USS *Enterprise* NCC 1701

*This book is dedicated to those who wish to explore
all aspects of how space and spatial relationships influence
the biology of species and ecosystems, but don't know
where to start.*

Table of Contents

	Page
Preface	ii
1. Introduction	1
2. How To Use The QGIS® 2.8.3 Software User Interface	8
3. Exercise One: Creating A High Quality Map Of Your Data In QGIS	10
4. Exercise Two: Creating Raster Data Layers Of Environmental Variables In QGIS	36
5. Exercise Three: Linking Spatial Data Sets In QGIS And Conducting Basic Descriptive Analyses Using R	63
6. Exercise Four: Conducting Linear Regressions With Generalised Linear Modelling (GLM) Using QGIS And R	85
7. Exercise Five: Conducting Non-Linear Regressions With Generalised Additive Modelling (GAM) Using QGIS And R	99
Appendix I: List Of Commands Containing The R Code Used Within This Workbook	115
Appendix II: Trouble-shooting Problems With Running Code In R	125
Appendix III: How To Transfer Data From R To QGIS	127

Preface

This workbook is a companion volume to *GIS For Biologists: A Practical Introduction For Undergraduates*. It is designed to augment the information on using GIS in biological research provided in that book. As a result, it assumes that you already have some level of familiarity with GIS. In addition, it is based around free-to-access, open source software. Specifically, it uses two such packages: QGIS for the GIS-based components and R for statistical analyses. Working with both of these packages, rather than just one or other, allows you to make the most of the specialist tools available within each of them.

The five exercises in this workbook demonstrate how to integrate QGIS and R to allow you to conduct high quality spatial analyses by accessing and combining the powerful mapping, data layer creation, editing and processing tools from QGIS with the equally powerful analytical tools from R. These exercises are based around data from a real biological field study and include: creating a GIS project to process your data and create a map suitable for publication; creating environmental raster data layers; linking environmental data to biological data and conducting basic descriptive analyses from the resulting data set; and running statistical analyses (GLMs and GAMs) to investigate spatial relationships in this combined data set. Working through these five exercises will help you gain experience in integrating QGIS and R for spatial analyses, and provide you with the confidence to apply these skills to your own research. These exercises are presented in the same easy-to-follow flow diagram-based format used in *GIS For Biologists: A Practical Introduction For Undergraduates*. They are accompanied by images which shows how your spatial analysis project should look as you progress through the exercises, allowing you to compare your own work to the expected results.

The exercises in this book build on exercise two (*How To Create Your Own Feature Data Layers*) from *GIS For Biologists: A Practical Introduction For Undergraduates* and use the same basic data sets. However, they do not require you to have completed that exercise first.

--- Chapter One ---

Introduction

The aim of this workbook is to help biologists expand their spatial analysis skills and familiarise themselves with the processes required to conduct a biologically meaningful spatial analysis project. To do this, it uses the same Task Oriented Learning (TOL) approach used in *GIS For Biologists: A Practical Introduction For Undergraduates* to provide five exercises based around the spatial processing and analysis of biological data, an increasingly important topic in many areas of biology. As such, it does not represent a stand-alone GIS book and is meant to act as a companion to the original book rather than to replace it in any way. Similarly, it does not provide any background information on using GIS as this has already been covered within *GIS For Biologists: A Practical Introduction For Undergraduates*. Instead, it simply provides instructions on how to do the exercises themselves. In addition, rather than relying on any commercial software packages, it is based around free-to-access, open source software. Specifically, it uses two such packages: QGIS for the GIS-based components and R for statistical analyses. In order to allow detailed and complete instructions to be given for each exercise, a single specific version of QGIS has been selected. This is QGIS 2.8.3 for Windows OS users or 2.8.4 for Mac OS users. While this is an older version of QGIS, it is the same version used for *GIS For Biologists: A Practical Introduction For Undergraduates*, and it provides a stable version that works well for these exercises. This version of QGIS can be downloaded from *www.GISinEcology.com/gis-for-biologists*. However, the instructions are provided in such a way that the information can be easily transferred to newer versions of QGIS (and, indeed, other GIS software packages).

The exercises provided in this workbook are designed to be worked through in a sequential manner. This is because the same data are used throughout these exercises and you will need to use some of the data layers and data sets generated in earlier exercises for later ones. In addition, each exercise represents a discrete, but important, step in the process of learning how to integrate QGIS and R to enable you to make the most of the tools provided by these two powerful open source software packages. For example, the first exercise covers how to add your biological data to a GIS project created in QGIS, process

Introduction

it and then create a map suitable for inclusion in a publication from these data, while later exercises cover the processing of environmental variables, linking data sets together, and conducting spatial analyses on these data in R. The data sets used in each exercise can be downloaded from *www.gisinecology.com/gis-for-biologists-workbooks*.

The exercises are provided using the same flow diagram-based approach used in *GIS For Biologists: A Practical Introduction for Undergraduates*. This means that for each exercise, you will first find an outline of what will be achieved by the end of it, why it is useful for biologists to be able to do this and what data you will need to start with. You will then find a series of numbered instruction sets which will take you through all the steps you need to do to complete that specific exercise. In order to allow you to know whether you are progressing correctly, figures are provided at regular intervals which will show you what the contents of the various windows of your GIS/statistical software should look like at that specific point in the exercise.

Before starting these exercises, it is worth taking the time to understand why it is beneficial to integrate QGIS and R (see figure 1 for a summary of these points), and the different ways that you can do this. QGIS is a software package designed for creating geographic information systems (GIS) and it has a wide range of powerful tools for creating, editing and processing spatial data. It also has powerful and dynamic data viewing and interrogation capabilities that allow you to quickly and easily visualise and assess the spatial distribution of your data. However, while it is well-suited for viewing, creating, processing and linking spatial data, as a GIS software package, it lacks the ability to do anything beyond the most basic statistical analyses. In contrast, R, as a dedicated analytical software package, has powerful data analysis capabilities, including the ability to apply a wide range of statistical techniques to spatial data. However, while it does have basic GIS capabilities, it lacks the ability to easily create, edit and process spatial data. In addition, R has very limited data viewing and interrogation capabilities, meaning it is difficult to use it to visualise and assess the spatial elements of your data in any sort of meaningful way. This is important when it comes to assessing the results of any tasks that involve spatial processing to ensure they have been carried out correctly and that they have successfully created the required outputs. From this brief comparison, it can be seen that these two software packages complement each other, with the powerful capabilities of one making up for the limitations in the other. This means that by integrating QGIS and R, you can create an incredibly powerful set of tools that can do almost anything that you would wish to be able to do when conducting spatial analyses of biological data.

QGIS	**R**
• No native statistical analyses.	• Powerful statistical analyses.
• Powerful (and growing) GIS functions.	• Basic GIS functions (but growing).
• Graphic User Interface-based.	• Code-based.
• Easy to check, define and transform projections.	• Difficult to deal with projections.
• Powerful data viewing and interrogation capabilities.	• Very limited data viewing capabilities.
• Easy to create and edit data layers.	• Very limited ability to create and edit data layers.
• Can easily open almost any spatial data format.	• Can be tricky to open some file formats.
What is QGIS good for?	**What is R good for?**
Viewing data, dealing with projections, checking compatibility, doing complex GIS functions, creating new data layers, reading unusual file formats.	Conducting statistical analyses and specialist GIS functions.

Figure 1. *A comparison of the advantages and disadvantages of using QGIS and R for spatial analyses, including identifying what each software package is good for. By integrating QGIS and R, you can make the most of the advantages of using each individual package to create a spatial analysis system that is more powerful than either component is on its own.*

While many powerful software packages do not necessarily work well together, this is not the case of QGIS and R. Due to their open source nature, it has been possible for their developers to create tools that allow you to easily move from one environment to the other and back again. For example, attribute tables, shapefiles and raster data layers created in QGIS can easily be imported into R (see examples provided in exercises three to five), while tables created in R can easily be exported and added to GIS projects in QGIS (see appendix III).

Introduction

The easiest way to integrate these two software packages is to use a single file storage system and a common format to transfer data from a GIS project created in QGIS into R, and to use this same common format to transfer data and the results of analyses conducted in R back into your GIS project in QGIS. This is done by using something called the 'Shapefile' approach to creating GIS projects. In this approach, a GIS project, and all the data layers it contains, are stored as separate files in a dedicated folder on the hard-drive of your computer. This same folder is then also used to store the R project which will use data from this GIS project. Any data that needs to be transferred between the two software packages can then be exported from one software packaged in a format that can be read in the other and accessed from this shared folder.

The second way to integrate QGIS and R is to directly link R to QGIS. This can either be done with a dedicated R portal in QGIS (versions prior to QGIS 3) or with an R plugin (QGIS 3 and onwards). These options allow you to run complex analyses on your data sets using R tools directly from within the main QGIS user interface. This means that you can seamlessly move between using tools from QGIS and R within a single user interface, and easily use QGIS's powerful data layer creation, editing, processing and viewing tools on data sets you wish to analyse in R.

This workbook will primarily assume that you are using the first way to integrate QGIS and R, and that you are using a single file storage system and common data formats for transforming data between the two software packages. This approach was selected as it provides a method that is applicable to all versions of QGIS and R, and it avoids a number of issues that are commonly encountered when linking the two packages together. All the R code required to complete the exercises in this workbook is provided in the accompanying flow diagrams, in appendix I, and in the text file called R_CODE_GIS_WORKBOOK_1.DOC which you can find in the compressed folder containing the data you will download at the start of exercise one. If you do not already have a copy of R on your computer, you will need to download and install it. You can do this by visiting *www.r-project.org*. If you are new to using R and wish to learn more about creating R codes to do specific tasks, we recommend reading *Getting Started with R: An Introduction for Biologists* by Andrew P. Beckerman and Owen L. Petchey.

NOTE: While this workbook contains information on how to run number of different statistical tests (e.g. a t-test, generalised linear modelling and generalised additive modelling), it is not a statistical textbook and these are only provided to illustrate how to transfer data sets from QGIS to R, and then conduct spatial analyses on them. This means that the statistical tests used in this book should not be applied to any other data set without first ensuring that they are appropriate. In addition, even if they are appropriate, this does not mean that other statistical tests are not just as valid for use with a particular data set or for a particular project. If you wish to learn more about selecting appropriate statistical tests for analysing your own data sets, and how to apply them in a correct and biological meaningful manner, you will need to consult one of the many good statistical textbooks which are available. In particular, we would recommend those written by Alain Zuur (see *www.highstat.com/books.htm* for more details).

NOTE: As with many things in GIS, there may be more than one way to do the processes required to complete the exercises outlined in this workbook. The instructions presented here will work for the data sets provided, and this means they should also work in most other circumstances. However, if you find an alternative way to do them which works for your data, or if you have someone who can show you how to do them in another way, feel free to do them differently.

--- Chapter Two ---

How To Use The QGIS 2.8.3 Software User Interface

In this workbook, QGIS software (also known as Quantum GIS) is used to illustrate how the steps outlined in each instruction set can be completed using this free GIS software package. Specifically, these instructions are for version 2.8.3 of QGIS. This version has been selected as it is very stable and it is the same version used for the exercises in *GIS For Biologists: A Practical Introduction For Undergraduates*. However, the same basic steps and tools also apply to more recent versions of QGIS. To download the required version of QGIS, please visit *www.GISinEcology.com/gis-for-biologists* and click on the links provided.

When you first open QGIS 2.8.3, you will find that the main user interface is divided into a number of sections and windows. However, before you start working through the exercises in this book, you will need to standardise its appearance. To do this, first click on the VIEW menu and select PANELS. Make sure that only the panels which are set to display are LAYERS and TOOLBOX. If any others are set to display, uncheck the box next to their names. Make sure that the panel titled LAYERS is on the left hand side and the one titled PROCESSING TOOLBOX is on the right. At the bottom of the PROCESSING TOOLBOX window, you have the option of selecting an ADVANCED INTERFACE or a SIMPLIFIED INTERFACE. Select the ADVANCED INTERFACE option. Next, you may find that you have one or more toolbar(s) running vertically down the left hand side of the QGIS window. If this is the case, click on the stippled bar at the top of each of one and drag it to the top of the window where the other toolbars are displayed. Finally, click on the VIEW menu again and select TOOLBARS. Make sure all the toolbars are set to display with the exception of ADVANCED DIGITIZING and DATABASE. The QGIS window should now look like figure 2.

The QGIS user interface has a number of key areas. Along the very top of the window is a MAIN MENU BAR which allows you to access tools for creating and saving GIS projects (under the PROJECT menu), for changing what is displayed in the QGIS interface (under the VIEW menu), as well as the various editing and geoprocessing tools (under the remaining menus). Below the MAIN MENU BAR is an OPTIONAL TOOLBARS AREA, where you can display toolbars for specific tool sets and plugins to allow you to access them quickly and easily. On the left hand side, there is a section called LAYERS which displays a list of all the data layers in your GIS project and it will be referred to as the TABLE OF CONTENTS window throughout this workbook. To the right of the TABLE OF CONTENTS window is the MAP window. This is where all the active data layers in your GIS project are displayed. On the right hand side is the PROCESSING TOOLBOX window. This window allows you to access a variety of tools (including those from R), and will be referred to as the TOOLBOX window. At the bottom is an X-Y COORDINATE DISPLAY area, which provides the X and Y coordinates for the position of the cursor in the MAP window, as well as tools for setting the scale of your MAP window, for deciding how your map is drawn and for setting the coordinate reference system (CRS). However, in this workbook, it will be referred to as the projection/coordinate system.

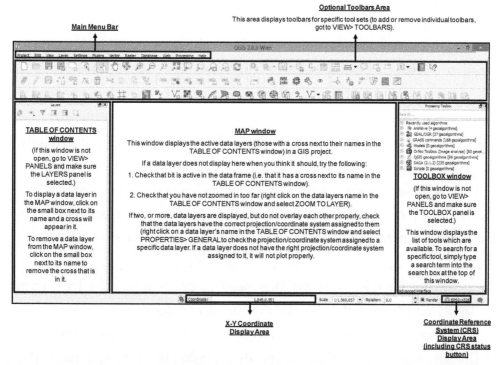

Figure 2. *The standard layout of the QGIS user interface which will be assumed for the exercises in this workbook. Within this book, the individual sections of the window will be referred to using the names which are <u>underlined</u> and in* **bold***.*

--- Chapter Three ---

Exercise One: Creating A High Quality Map Of Your Data In QGIS

For the exercises in this workbook, you will be working with data from a study area in central Scotland where researchers from the Scottish Centre for Ecology and the Natural Environment (SCENE) field station (see *tinyurl.com/GFB-Link20* for more information on this facility) have been looking at factors which influence the breeding success of hole-nesting birds in an area of native oak woodland. As the exercises progress, you will learn how to create a data layer of the nest box locations from data held in a spreadsheet, join it to information on nest box occupancy by difference species, link the nest box locations to local environmental information, such as land elevation, and then examine the relationships between nest box occupancy, breeding success and a variety of environmental variables.

However, as with most research projects, the first thing you will do with these data is to create a map. This map will show not only the locations of each nest box, but also which nest boxes were occupied by blue tits (one of most common hole-nesting bird species in Scotland) during a particular breeding season. It will also contain additional information to help put the nest box data in a wider context, such as the limits of the patch of native oak woodland that nest boxes are sited in, and local environmental and human landscape features. As a GIS software package, QGIS is well-suited to creating such high quality maps, and it has much greater and more flexible map-making capabilities than R. This means that for this exercise, you will be using QGIS to create your map.

Before you start this exercise, you will first need to create a new folder on your C: drive called QGIS_R_WORKBOOK. To do this on a computer with a Windows operating system, open Windows Explorer and navigate to your C:\ drive (this may be called Windows C:). To create a new folder on this drive, right click on the window displaying the contents of your C:\ drive and select NEW> FOLDER. Now call this folder QGIS_R_WORKBOOK by typing this into the folder name to replace what it is currently called (which will most likely be NEW FOLDER). This folder, which has the address C:\QGIS_R_WORKBOOK, will be used to store all files and data for the exercises in this book.

Next, you need to download the source files for six existing data layers from *www.gisinecology.com/gis-for-biologists-workbooks/*. When you download the required files, save them into the folder C:\QGIS_R_WORKBOOK which you have just created. These data layers are:

1. SCENE_Oak_Woodland.shp: This is a polygon data layer which has a single polygon representing an area of native oak woodland in central Scotland where a number of nest boxes have been sited to study the breeding behaviour of hole-nesting birds. It was created specifically for this research project. It is in the British National Grid projection and is based on the OSGB 1936 datum.

2. SCENE_Roads.shp: This is a line data layer which represents sections of road near the SCENE field station. It was obtained from the OS OpenData Meridian 2 data set (see *tinyurl.com/GFB-Link6* for more information). It is in the British National Grid projection and is based on the OSGB 1936 datum.

3. SCENE_Loch_Lomond_Shoreline.shp: This is a line data layer which is a high resolution representation of the shoreline of Loch Lomond. This is a large freshwater lake and the oak woodland study area stands on its shore. This data layer was created specifically for this research project. It is in the British National Grid projection and is based on the OSGB 1936 datum.

4. SCENE_Buildings.shp: This is a point data layer that shows the location of the buildings that make up the SCENE field station. It is in the British National Grid projection and is based on the OSGB 1936 datum.

5. Dubh_Loch.shp: This is a polygon data layer which has a single polygon representing a small body of freshwater on the edge of the oak woodland around the SCENE field station. It is in the British National Grid projection and is based on the OSGB 1936 datum

6. Forestry_Track.shp: This is a line data layer which represents a track through the oak woodland which could act as a source of disturbance for the hole-nesting birds within this area, and so could potentially influence breeding success. It is in the British National Grid projection and is based on the WGS 1984 datum OSGB 1936 datum.

NOTE: The data layers for the SCENE buildings, the Dubh Loch and the forestry track were created as part of exercise two in *GIS For Biologists: A Practical Introduction For Undergraduates*. That exercise shows you a number of different ways that you can create your own feature data layers, including drawing them directly in QGIS, tracing a feature from Google Earth and importing data recorded on a GPS receiver.

Once you have all these files downloaded into the correct folder on your computer, and you understand what is contained within each file, you can move on to starting the GIS project you will use for the exercises in this book. The starting point for this is a blank GIS project. To create a blank GIS project, first open QGIS. Once it is open, click on the PROJECT menu and select SAVE AS. In the window which opens, save your GIS project as EXERCISE_ONE in the folder called C:\QGIS_R_WORKBOOK.

You are now ready to begin step one of this exercise. Instructions on how to do this step start on the next page. As you work through the instructions for this exercise, you can compare your own work to the images of the contents of the MAP window, the TABLE OF CONTENTS window and/or the TABLE window that are provided to allow you to assess whether you have successfully completed each step before you move on to the next one.

STEP 1: SET THE PROJECTION AND COORDINATE SYSTEM OF YOUR DATA FRAME:

The selection of an appropriate projection/coordinate system is a critical first step in any GIS project, and it is important that you select one that is appropriate for the area of the world you are working in, and the size of your study area. If you select an inappropriate projection/coordinate system, you may find that you cannot create accurate raster data layers of environmental variables, such as land elevation or slope. This means you need to think carefully about what projection/coordinate system will be appropriate for your particular study before you starting your GIS project. You can find advice on how to select an appropriate projection/coordinate system for a specific project in chapter four of *GIS For Biologists: A Practical Introduction For Undergraduates*. Once you have selected an appropriate projection/coordinate system, you need to set your GIS project to use it before you do anything else (including adding any data to it). This helps ensure that you do not, at any point, accidently use the wrong one. This is important as mix-ups with the projection/coordinate systems account for about 90% of the problems you are likely to encounter with the spatial processing required for conducting meaningful analyses of biological data.

For this exercise, you will use a pre-existing projection/coordinate system called British National Grid. This is a transverse mercator projection which is specifically designed to minimise distortion of features in the British Isles and it uses the OSGB 1936 datum. This projection/coordinate system is used because it is the same as the one used during data collection and for the existing data layers which you will use during the exercises in this workbook. To set your data frame to use the British National Grid projection/coordinate system, work through the flow diagram provided on the next page.

NOTE: Within QGIS, the projection/coordinate system is referred to as the coordinate reference system or CRS.

Exercise One: Creating A High Quality Map Of Your Data In QGIS

Data frame with no associated projection and coordinate system

When you open a new GIS project, there will be an empty data frame. It will not have a projection or coordinate system associated with it.

In the main QGIS window, click on the CRS STATUS button in the bottom right hand corner (it is rectangular with a dark circular design on it beside the text EPSG: 4326). This will open the PROJECT PROPERTIES CRS window. Click on the box next to ENABLE 'ON THE FLY' CRS TRANSFORMATION so that a cross appears in it. This allows you to select a specific projection/coordinate system for your data frame. It also allows you to have data layers in different projection/coordinate systems in your GIS project and still have them plot in the correct places relative to each other. For biological GIS projects, you will almost always want to select this option.

1. Set your data frame to use a pre-existing projection/coordinate system

Next, type the name of the projection/coordinate system you want to use into the FILTER section (in this case it will be BRITISH NATIONAL GRID). In the COORDINATE REFERENCE SYSTEMS OF THE WORLD section, click on the OSGB 1936 / BRITISH NATIONAL GRID option with the authority EPSG: 27700 under PROJECTED COORDINATE SYSTEMS> TRANSVERSE MERCATOR. This will add this projection/coordinate system to the SELECTED CRS section further down the window. Now click OK to close the PROJECT PROPERTIES CRS window. **NOTE:** You may have to reduce the size of this window to be able to see the OK button.

2. Check that the projection/coordinate system you have selected is appropriate

Once you have selected a projection/coordinate system, you need to check that it is appropriate. This involves examining how data layers look in it. For this exercise, this will be done in the next step by adding a series of existing data layers and checking that the features in them plot in the expected places and that any polygons have the expected shape.

Data frame set to use a pre-existing projection/coordinate system

To check that you have done this step correctly, click on PROJECT on the main menu bar and select PROJECT PROPERTIES. In the PROJECT PROPERTIES window which will open, click on the CRS tab on the left hand side and make sure that the SELECTED CRS section has the following text in it:

OSGB 1936/ British National Grid

Underneath this, it should say:

+proj=tmerc +lat_0=49 +lon_0=-2 +k=0.9996012717 +x_0=400000 +_0=-100000 +ellps=airy +towgs84=446.448,-125.157,542.06,0.15,0.247,0.842,-20.489 +units=m +no_defs

This is known as the Proj.4 string, which is used to define the characteristics of the projection/coordinate system.

If it does, click OK to close the PROJECT PROPERTIES window. If it does not, you will need to repeat this step until you have assigned the correct projection/coordinate system to your data frame. Once you have successfully completed this step, click on the PROJECT menu on the main menu bar and select SAVE to save the changes you have made to your GIS project.

STEP 2: ADD EXISTING DATA LAYERS YOU WISH TO DISPLAY ON YOUR MAP TO YOUR GIS PROJECT:

Once you have set the projection/coordinate system of your data frame, you are ready to add some existing data layers to it. For this exercise, you will use a number of existing data layers which provide information about a variety of different features of the native oak woodland study area where the nest boxes are sited. This will include the area of the woodland itself (SCENE_OAK_WOODLAND), information about the position of the edge of a large lake (LOCH_LOMOND_SHORELINE), the outline of a small body of water that borders the oak woodland (DUBH_LOCH) and a number of human landscape

Exercise One: Creating A High Quality Map Of Your Data In QGIS

features. These include the location of nearby roads (SCENE_ROADS), a track through the woodland (FORESTRY_TRACK) and the location of the buildings that make up the SCENE field station (SCENE_BUILDINGS). These existing data layers are all in the British National Grid projection which is being used for this exercise. To add these data layers to your GIS project, work through the following flow diagram:

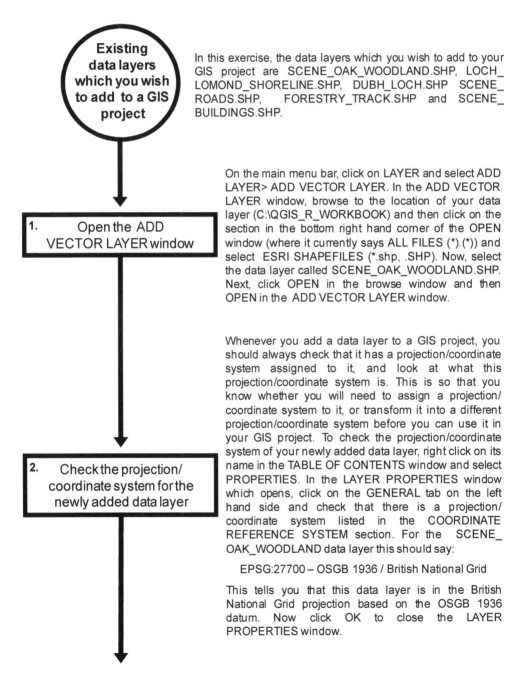

Exercise One: Creating A High Quality Map Of Your Data In QGIS

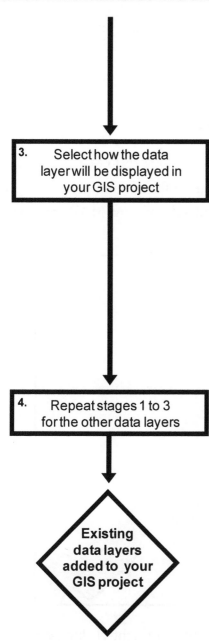

Right click on the name of the SCENE_OAK_ WOODLAND data layer in the TABLE OF CONTENTS window, and select PROPERTIES. In the LAYER PROPERTIES window, click on the STYLE tab on the left hand side. In the top left hand corner of the STYLE tab, select SINGLE SYMBOL from the drop down menu, and then click on FILL> SIMPLE FILL in the section below it. Now, on the right hand side of the window, click on the box next to COLORS FILL. In the SELECT FILL COLOR window which will open, select a green shade from the colour selector on the left hand side and then click OK. Now, click on the box next to BORDER and select a dark green shade and then click OK to close the SELECT COLOR window. Next enter 1.00 for BORDER WIDTH and then click OK to close the LAYER PROPERTIES window. You will see that the way the SCENE_OAK_WOODLAND data layer is displayed in the MAP window has now changed to the new settings you have just selected.

Repeat stages 1 to 3 for the data layers called LOCH_ LOMOND_SHORELINE, DUBH_LOCH, SCENE_ ROADS, FORESTRY_TRACK and SCENE_ BUILDINGS (in this order) to add them to your GIS project. For the line data layers, select LINE> SIMPLE LINE, and then a PEN WIDTH of 1.0, but select black for the colour of the shoreline and a grey shade for the roads and the forestry track. For the DUBH_LOCH layer, use light blue for the fill and dark blue for the border (making sure to set the border width to 1.00). Finally, for the SCENE_BUILDINGS layer, set it to display as a black square symbol of size 2.

At the end of this step, your TABLE OF CONTENTS window should look like the image at the top of the next page.

Exercise One: Creating A High Quality Map Of Your Data In QGIS

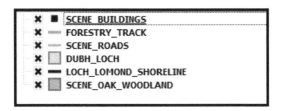

While the contents of your MAP window should look like this (**NOTE:** Depending on the shape of your screen, you may see additional information from the LOCH_LOMOND_SHORELINE data layer on the left hand side of your MAP window – this is okay):

If the contents of your MAP window do not look like this, try right-clicking on the name of the SCENE_OAK_WOODLAND data layer in the TABLE OF CONTENTS window and selecting ZOOM TO LAYER. If it still does not look right, remove all the data layers from your GIS project by right-clicking on their names in the TABLE OF CONTENTS window and selecting REMOVE. Next, go back to step one and ensure that you have set the projection/coordinate system of your data frame correctly, and then repeat step two. Once you have successfully completed this step, click on the PROJECT menu on the main menu bar and select SAVE to save the changes you have made to your GIS project.

STEP 3: CREATE A NEW POINT DATA LAYER OF THE NEST BOX LOCATIONS FROM A LIST OF COORDINATES IN A SPREADSHEET:

Once you have added the required existing data layers to your GIS project, you can create a new point data layer showing the locations of the nest boxes being used to study breeding success in hole-nesting birds from a list of coordinates held in a spreadsheet. This will be done in two parts. Firstly, you will plot the locations of all the nest boxes in the spreadsheet and make them into a point data layer. Once you have done this, you will select only those nest boxes which are within the oak woodland study area around the SCENE field station and make a new data layer with just these locations in it.

NOTE: Before you start this exercise, you need to ensure that the SPATIAL QUERY plugin is installed and activated in your version of QGIS. To do this, click on PLUGINS on the main menu bar, and select MANAGE AND INSTALL PLUGINS (you will need to have access to the internet to be able to manage your plugins using this option). In the PLUGINS window which opens, click on the ALL tab on the left hand side and then type SPATIAL QUERY into the SEARCH section of this window. If the SPATIAL QUERY PLUGIN is not installed, select it and click on the INSTALL PLUGIN button. If it is installed, check that there is a cross (or a tick) in the white box next to its name (this means that it has been activated). If there is no cross or tick in this box, click on it so that a cross/tick appears in it. You can now click CLOSE to close the PLUGINS window.

To create your new point data layer of the nest box locations from a list of coordinates in a spreadsheet, work through the flow diagram that starts at the top of the next page.

Exercise One: Creating A High Quality Map Of Your Data In QGIS

Locational data in a delimited text file with latitude and longitude coordinates

The locational data for nest boxes are in the file C:\QGIS_R_WORKBOOK\NESTBOX_LOCATIONS.TXT.

1. Plot locational data in your GIS project

Click on LAYER on the main menu bar and select ADD LAYER> ADD DELIMITED TEXT LAYER. Once the window for this tool is open, you need to browse to the file which contains the data you wish to plot by clicking on the browse button at the right hand end of the FILE NAME section of the tool window. In this case, browse to the folder called C:\QGIS_R_WORKBOOK and select the file named NESTBOX_LOCATIONS.TXT before clicking on the OPEN button. In the LAYER NAME section of the tool window, enter NESTBOX_LOCATIONS_V1 (this is the name which will be displayed in the TABLE OF CONTENTS window for this layer). For FILE FORMAT select CUSTOM DELIMITERS, and then, directly below this, select TAB. For RECORD OPTIONS, make sure that the FIRST RECORD HAS FIELD NAMES option is selected. For GEOMETRY DEFINITION, select POINT COORDINATES, and then make sure that LONGITUDE is selected for X FIELD and that LATITUDE is selected for Y FIELD. Finally, click OK to create a point data layer from your data. This will cause a warning to appear indicating that the CRS has not been set. This is okay. You will select the appropriate projection/coordinate system for your data in the next stage of this step.

2. Assign the correct projection/coordinate system to the data layer

Right click on the data layer called NESTBOX_LOCATIONS_V1 in the TABLE OF CONTENTS and select PROPERTIES. In the LAYER PROPERTIES window, click on the GENERAL tab and then click on the SELECT CRS button at the right hand end of the COORDINATE REFERENCE SYSTEM section (it has a picture of a globe and small yellow square on it). In the COORDINATE REFERENCE SYSTEM SELECTOR window which opens, type WGS 84 into the FILTER section (making sure that you leave a space between the letters WGS and the number 84). Next, in the COORDINATE REFERENCE SYSTEMS OF THE WORLD section, select WGS 84 under GEOGRAPHIC COORDINATE SYSTEM. **NOTE:** If your coordinates were in a different projection/coordinate system, you would set this different system at this stage. In the SELECTED CRS section, you should now see WGS 84, and in the window below it the following proj.4 string:

+proj=longlat +datum=WGS84 +no_defs

Finally, click OK to set this as the projection/coordinate system for the point data layer you have just created, and then click OK to close the LAYER PROPERTIES window.

In order to make a permanent version of your data layer, you need to convert it into a shapefile. This can be done using the SAVE AS tool. To access this tool, right click on the name of the data layer you just created in the TABLE OF CONTENTS window, and select SAVE AS. This opens the SAVE VECTOR LAYER AS. In this window, for FORMAT select ESRI SHAPEFILE. Next, enter the following address and file name into the SAVE AS section of the window: C:/QGIS_R_WORKBOOK/NESTBOX_LOCATIONS_V2.SHP. **NOTE:** This address uses slashes (/) rather than the usual backslashes (\). In the CRS section, make sure that the option starting with PROJECT CRS is selected. Now, select ADD SAVED FILE TO MAP and then click on the OK button.

3. Transform your data layer into the projection/coordinate system being used for your GIS project and save it as a shapefile

Finally, right click on the name of the data layer you created in stages 1 and 2 (NESTBOX_LOCATIONS_V1) in the TABLE OF CONTENTS window and select REMOVE from the menu which appears to remove it from your GIS project. This is because you no longer need it now that you have transformed your data layer into the projection/coordinate system you are using for your project and have converted it into a shapefile.

If you right click on the name of the data layer you have just created (NESTBOX_LOCATIONS_V2) in the TABLE OF CONTENTS window and select ZOOM TO LAYER, you will see that there are two groups of nest boxes in it, one set in the SCENE_OAK_WOODLAND polygon and one further south. You now want to select those nest boxes in the SCENE oak woodland and make a new data layer with just these locations in it. To do this, click on VECTOR on the main menu bar and select SPATIAL QUERY> SPATIAL QUERY (**NOTE:** if this option is not available, check that the SPATIAL QUERY plugin has been activated – see note on page 17 for more information). This will open the SPATIAL QUERY window. For SELECT SOURCE FEATURES FROM, select NESTBOX_LOCATIONS_V2 from the drop down menu. For REFERENCE FEATURES OF, select SCENE_OAK_WOODLAND, and then for WHERE THE FEATURE, select WITHIN. Now click APPLY to run the spatial query and then CLOSE to close the SPATIAL QUERY tool window.

4. Make a new data layer by selecting a subset of data in a data layer

To make a new data layer containing only the selected locations, right click on the data layer called NESTBOX_LOCATIONS_V2 in the TABLE OF CONTENTS window and select SAVE AS. This opens the SAVE VECTOR LAYER AS window. In this window, for FORMAT select ESRI SHAPEFILE. Next, enter the following address and file name into the SAVE AS section of the window: C:QGIS_R_WORKBOOK/SCENE_NESTBOX_LOCATIONS.SHP. In the CRS section, make sure that the option starting with LAYER CRS is selected. Now, select SAVE ONLY SELECTED FEATURES and then select ADD SAVED FILE TO MAP. Finally, to save the selected records as a new data layer, click on the OK button.

Exercise One: Creating A High Quality Map Of Your Data In QGIS

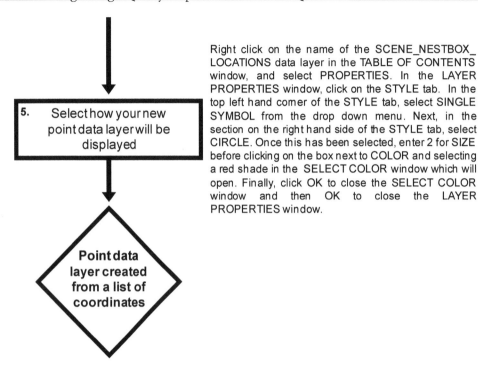

At the end of this step, your TABLE OF CONTENTS window should look like this:

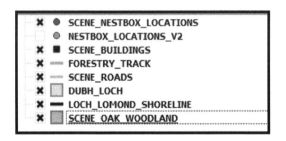

Click on the box next to the name of the data layer SCENE_NESTBOX_LOCATIONS in the TABLE OF CONTENTS window so that it is no longer displayed in the MAP window. Next, click on VIEW on the main menu bar and select SELECT> DESELECT FEATURES FROM ALL LAYERS. Finally, right click on the data layer called NESTOX_LOCATIONS_V2 in the TABLE OF CONTENTS window and select ZOOLM TO LAYER. The contents of your MAP window should now the image at the top of the next page.

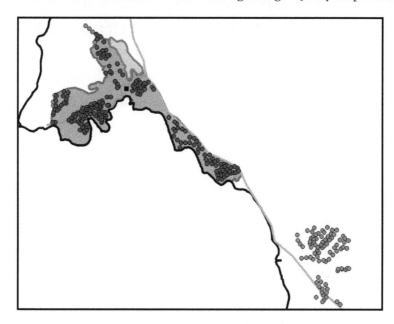

Now click on the box next to SCENE_NESTBOX_LOCATIONS in the TABLE OF CONTENTS window again (so that it is displayed in the MAP window once more) and uncheck the box next to NESTBOX_ LOCATIONS_V2 (so that it is no longer displayed in the MAP window). You will see that it contains only those nest box locations in the SCENE oak woodland. Right click on SCENE_OAK_WOODLAND in the TABLE OF CONTENTS window and select ZOOM TO LAYER. The contents of your MAP window should now look like this:

Exercise One: Creating A High Quality Map Of Your Data In QGIS

You can now remove the data layer NESTBOX_LOCATIONS_V2 from your project. To do this, right click on its name in the TABLE OF CONTENTS window and select REMOVE. Once you have successfully completed this step, click on the PROJECT menu on the main menu bar and select SAVE to save the changes you have made to your GIS project.

STEP 4: JOIN INFORMATION ABOUT WHICH NEST BOXES WERE OCCUPIED BY THE TARGET SPECIES TO THE NEST BOX DATA LAYER AND USE IT TO DISPLAY THIS INFORMATION IN THE MAP WINDOW:

At the moment, the point data layer showing the locations of the nest boxes (called SCENE_NESTBOX_LOCATIONS) does not contain any information about which boxes were occupied by the target species (in this case blue tits) and which were not for the breeding season under investigation. This information is contained in a spreadsheet file called NEST_BOX_BREEDING_DATA.XLS that can be found in the compressed folder downloaded at the start this exercise. Therefore, before a map can be created showing this information, it first has to be added to the attribute table of SCENE_ NESTBOX_LOCATIONS data layer. This is done using a table join to join the information on nest box occupancy to this attribute table based on the nest box identification number, which is found in both data sets. Once this has been done, it can be used to change the way the SCENE_NESTBOX_LOCATIONS data layer is displayed in the MAP window to show the required information. To do this, work through the flow diagram that starts at the top of the next page.

Exercise One: Creating A High Quality Map Of Your Data In QGIS

Point data layer of nest box locations and spreadsheet with information on species occupancy

The point data layer of the nest box locations is called SCENE_NESTBOX_LOCATIONS, while the spreadsheet file with the information about species occupancy is called NEST_BOX_BREEDING_DATA.XLS.

1. Add the spreadsheet with species occupancy information as a table to your GIS project

To add a spreadsheet to your GIS project as a table, you first need to install the SPREADSHEET LAYERS PLUGIN. To do this, click on PLUGINS on the main menu bar and select MANAGE AND INSTALL PLUGINS. In the window that opens, type SPREADSHEET into the search bar at the top and then select the SPREADSHEET LAYERS option once it appears. If it is not already installed, click INSTALL PLUGIN to install it. After it has been installed, check that it has been activated (there should be a tick or a cross in the box next to its name). Once this plugin has been installed and activated, close the PLUGINS window. Now click on PROJECT on the main menu bar and select OPEN RECENT> C:/QGIS_R_WORKBOOK / EXERCISE_ONE.QGIS.

Next, click on LAYER on the main menu bar and select ADD LAYER> ADD SPREADSHEET LAYER (this option has been added because the SPREADHSEET LAYERS plugin has been installed and activated). In the window that opens, click on the BROWSE button at the end of the FILE NAME section. Select the file called NEST_BOX_BREEDING_DATA.XLS from the folder C:/QGIS_R_WORKBOOK and click on the OPEN button. **NOTE:** If no data appear when you do this, try using the NEST_BOX_BREEDING_DATA.XLSX file instead. Set the LAYER NAME to NEST_BOX_ BREEDING_DATA and make sure that SHEET1 is selected in the SHEET section. Finally, make sure that there is no cross in the box next to HEADER AT FIRST LINE and then click OK to add this table to your GIS project.

Exercise One: Creating A High Quality Map Of Your Data In QGIS

2. Join the species occupancy data to the attribute table of the nest box locations point data layer

Right click on the data layer called SCENE_NESTBOX_LOCATIONS in the TABLE OF CONTENTS window and select PROPERTIES. In the LAYER PROPERTIES window, click on the JOINS tab and then click on the ADD JOIN button (it is at the bottom of the tab and has a green cross on it). In the ADD VECTOR JOIN window that opens, select NEST_BOX_BREEDING_DATA as the JOIN LAYER, and select NEST_BOX as the JOIN FIELD. For TARGET FIELD, select BOX_NUMBER. This will join the data from the NEST_BOX_BREEDING_DATA table to the attribute table for SCENE_NESTBOX_LOCATIONS data layer based on the contents of these fields. Next, click on the box next to CHOOSE WHICH FIELDS ARE JOINED and select only the field called SPECIES. This contains the information about which species occupied each nest box during the breeding season being examined. Now click on the box next to CUSTOM FIELD NAME PREFIX, enter BD_ (which stands for Breeding Data) in this section. Finally, click OK to make the join, and then click OK to close the LAYER PROPERTIES window.

3. Create a new field called BT_OCC with a 1 for nest boxes occupied by the target species (blue tits)

In order to easily identify which nest boxes were occupied by the target species (blue tits) in the breeding season under investigation, and which were not (even if they were occupied by another species), you need to create a new field that contains this information. To do this, first click on the data layer called SCENE_NESTBOX_LOCATIONS in the TABLE OF CONTENTS window and select OPEN ATTRIBUTE TABLE. In the TABLE window that opens, click on the TOGGLE EDITING button in the top left hand corner (it has a picture of a pencil on it), and then click on the SELECT FEATURES USING AN EXPRESSION button (it has a yellow square and symbol that looks like the letter E on it). In the SELECT BY EXPRESSION window that opens, enter the following expression "BD_SPECIES" = 'BT' and then click SELECT. This will select all the nest boxes occupied by blue tits (during the original data entry, these have been coded with the initials BT). Now, click CLOSE to close the SELECT BY EXPRESSION window and then click the OPEN FIELD CALCULATOR button at the top of the TABLE window (it as a picture of an abacus on it). In the FIELD CALCULATOR window, select ONLY UPDATE 66 SELECTED FEATURES at the top, and then CREATE A NEW FIELD below this. For OUTPUT FIELD NAME, enter BT_OCC, and for OUTPUT FIELD TYPE select WHOLE NUMBER (INTEGER). In the EXPRESSION window, enter the value 1 and then click OK. This will create a field called BT_OCC and fill with a value of 1 for all the nest boxes occupied by blue tits.

Click on the INVERT SELECTION button at the top of the TABLE window (it as two green arrows on a yellow triangle on it). This will select all the nest boxes not occupied by blue tits (even if there were occupied by another species – this is what you want to do as you are currently only interested in blue tits). Once you have done this, click the OPEN FIELD CALCULATOR button at the top of the TABLE window (it as a picture of an abacus on it). In the FIELD CALCULATOR window, select ONLY UPDATE 132 SELECTED FEATURES at the top, and then UPDATE EXISTING FIELD below this. Select BT_OCC from the drop down menu directly below this option to select this field to update. In the EXPRESSION window, enter the value 0 and then click OK. This will fill in a value of 0 for all the nest boxes not occupied by blue tits. You can now click on the SAVE EDITS button at the top of the TABLE window (it has a picture of an old-fashioned diskette on it), and then click on the UNSELECT ALL button (it is next to the SELECT BY EXPRESSION button). Finally, click on the TOGGLE EDITING MODE button to close the editing session before closing the TABLE window.

4. Add a 0 to the field called BT_OCC for all nest boxes not occupied by blue tits

Right click on the name of the SCENE_NESTBOX_ LOCATIONS data layer in the TABLE OF CONTENTS window, and select PROPERTIES. In the LAYER PROPERTIES window, click on the STYLE tab. In the top left hand corner of the STYLE tab, select CATEGORIZED from the drop down menu. Next, select BT_OCC for COLUMN and then click on the CLASSIFY button. This will add three options to the central part of this STYLE tab window. Select a red colour and a symbol size of 2 for the symbol with a value of 0 (these are nest boxes not occupied by blue tits in this data set). This is done by double-clicking on the existing symbol for that category to open the SYMBOL SEELCTOR window. Next, select a blue color and a symbol size of 3 for the symbol with a value of 1 (these are nest boxes that were occupied by blue tits). Now, click on the symbol with no number next to it and click the DELETE button. Finally, click OK to close the LAYER PROPERTIES window.

5. Set your nest box location data layer to display blue tit nest box occupancy

Blue tit nest box occupancy identified and displayed in your GIS project

25

Exercise One: Creating A High Quality Map Of Your Data In QGIS

When you have completed this step, the attribute table for the SCENE_NESTBOX_LOCATIONS data layer will look like the table below, with a temporary join field called BD_SPECIES and a new permanent field called BT_OCC added to it. If the attribute table is not open, right click on the name SCENE_NESTBOX_LOCATIONS in the TABLE OF CONTENTS window and select OPEN ATTRIBUTE TABLE.

	Box_Number	Latitude	Longitude	BT_OCC	BD_Species
0	138	56.12643700000…	-4.61782100000…	0	NULL
1	139	56.12641253999…	-4.61816079000…	0	NULL
2	141	56.12621899999…	-4.61796500000…	0	NULL
3	144	56.12598299999…	-4.61698600000…	0	GT
4	143	56.12606799999…	-4.61688500000…	0	NULL
5	142	56.12621599999…	-4.61699800000…	1	BT
6	137	56.12642499999…	-4.61700800000…	0	NULL
7	33	56.13024099999…	-4.61416500000…	0	NULL
8	30	56.13059400000…	-4.61469500000…	0	NULL
9	24	56.13129000000…	-4.61525100000…	1	BT
10	48	56.12906499999…	-4.61481100000…	1	BT
11	13	56.13031399999…	-4.61620800000…	0	NULL
12	14	56.13051600000…	-4.61645000000…	0	NULL
13	19	56.13106500000…	-4.61664700000…	0	GT
14	301	56.13180899999…	-4.61724900000…	1	BT
15	44	56.13033200000…	-4.61723800000…	1	BT
16	4	56.12951600000…	-4.61551600000…	0	NULL
17	3	56.12926999999…	-4.61511000000…	1	BT
18	300	56.13213900000…	-4.61671900000…	1	BT
19	302	56.13208499999…	-4.61702100000…	0	NULL
20	308	56.13259999999…	-4.61738300000…	1	BT

Now close the attribute table for the SCENE_NESTBOX_LOCATIONS data layer. Your TABLE OF CONTENTS window should look like this:

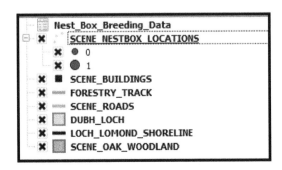

While the contents of your MAP window should look like this:

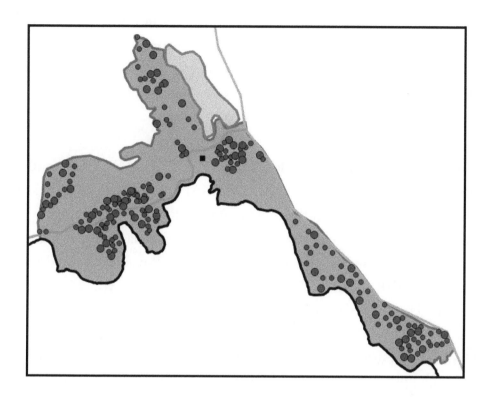

STEP 5: CREATE AND EXPORT A HIGH QUALITY MAP SHOWING NEST BOX OCCUPANCY BY THE TARGET SPECIES:

Now that you have identified which nest boxes were occupied by blue tits and which were not during the breeding season being investigated, you are now ready to create a high quality map showing this information. In QGIS, maps are created in the PRINT COMPOSER window. To open the PRINT COMPOSER window, click on PROJECT on the main menu bar and select NEW PRINT COMPOSER. A COMPOSER TITLE window will appear where you can enter a name for your map. Type in the name BLUE TIT OCCUPANCY MAP and then click OK. When the PRINT COMPOSER window first opens, it will be blank (this is okay) and it will be titled BLUE TIT OCCUPANCY MAP. The first thing you will do is set the paper size and orientation. To do this, click on

Exercise One: Creating A High Quality Map Of Your Data In QGIS

the COMPOSITION tab on the right hand side of the PRINT COMPOSER window, then under PAPER AND QUALITY select A4 for PAPER SIZE, MM for UNITS, PORTRAIT for ORIENTATION and 300 DPI for EXPORT RESOLUTION (this value will define the resolution of your final map, and it is this value you will need to change if you want to alter your map's resolution). Leave all the other sections with their default settings. **NOTE:** If the COMPOSITION tab is not visible, click on VIEW on the main menu bar and select PANELS> COMPOSITION.

To add the contents of your MAP window to the PRINT COMPOSER window, click on LAYOUT on its main menu bar and select ADD MAP. Now move your cursor to the top left hand corner of the white box in the middle of the PRINT COMPOSER window. Hold the left hand button of your mouse down, move the cursor diagonally to the bottom right hand corner and then release it. The contents of your MAP window should now appear in the PRINT COMPOSER window. When you initially look at it, you will undoubtedly not be too impressed. However, with a few simple steps, you can make it look much better.

The first thing that you will need to do is set the size and the position of your map. To do this, click on the ITEM PROPERTIES tab of the right hand section of the PRINT COMPOSER window. Here, you can change various settings for your map. In this section, scroll down to the POSITION AND SIZE section and click on the black arrow to open the position and size options. For PAGE, enter 1, and then for POSITION, enter 30 for X and 75 for Y, and you will see the position of your map change. Now, enter 150 for WIDTH and 125 for HEIGHT. Next, you will need to set the extent so that your map only shows the area you want it to. In the ITEM PROPERTIES tab, scroll up until you find the section called EXTENTS, and enter 236700 for X MIN, 694700 for Y MIN, 239100 for X MAX and 696700 for Y MAX. **NOTE:** If you did not already have these coordinates, you could get an idea of what coordinates would be appropriate to define the extent for a specific map before you open the PRINT COMPOSER window by going to the MAP window and using the ZOOM IN or ZOOM OUT tools until you can see all the data you wish to show on your map. Next, move the cursor to the bottom left of the area you wish your map to show. You can then read off the coordinates which you would need to use to set the left, or X MIN, (the first coordinate) and bottom, or Y MIN, (the second coordinate) limits for the extent of your map by looking at the coordinate display area at the bottom of the QGIS user interface (see figure 2 on page 7). Repeat this for the top right hand corner of the area you wish your map to show and read off the right, or X MAX, (the first coordinate) and top, or Y MAX, (the second coordinate) limits for the extent of you map. Once you have these values, you can set the extent of your map as outlined above.

The contents of your PRINT COMPOSER window should now look like this (**NOTE:** You may have to scroll the window with the map in it up or down to see it properly):

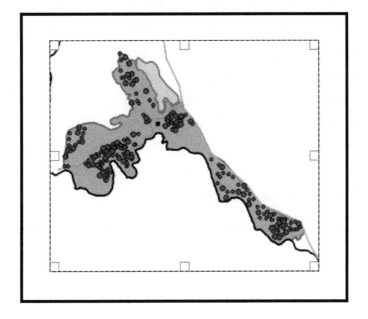

If it does not look like this, go back and check that you set the extent, size and position to the values given in the instructions before carrying on.

The next thing to do is to add a grid to your map so that people will know what part of the world it represents. As with the extent, size and position, this is also done in the ITEM PROPERTIES tab. In this tab, click on the black arrow beside GRIDS to reveal the GRID options, and then click on the + button to create your grid. Now, scroll down and click on the button with CHANGE on it next to CRS. This will open the COORDINATE REFERENCE SYSTEM SELECTOR window. In this window, enter WGS 84 into the FILTER section (making sure that you leave a space between the letters WGS and the number 84), and then in the COORDINATE REFERENCE SYSTEMS OF THE WORLD section, under GEOGRAPHIC COORDINATE SYSTEMS, select the WGS 84 with the AUTHORITY ID of EPSG: 4326 (**NOTE:** This is not a projected coordinate system, but rather a geographic coordinate system), and click OK. For interval, enter 0.01 for X and 0.01 for Y. A grid will now appear on the map you are creating. Next, click on the box beside LINE STYLE. This will open the SYMBOL SELECTOR window. In this window, select LINE> SIMPLE LINE and then select NO PEN from the drop down menu next to PEN STYLE, before clicking OK. This will change the grid so that it is no longer visible. Under GRID FRAME, select EXTERIOR TICKS for FRAME STYLE. Next, click on the box next to DRAW COORDINATES. This will activate the COORDINATES options. For FORMAT, select DEGREE, MINUTE WITH SUFFIX, and then for LEFT and RIGHT change the orientation setting from HORIZONTAL to VERTICAL ASCENDING. Next, click on the box beside FONT. In the SELECT FONT window which will open, select TIMES NEW ROMAN for the FONT and 12 for the SIZE, and then click OK. Now, enter 2 for DISTANCE FROM MAP FRAME and 1 for COORDINATE PRECISION. Finally, scroll down and click on the box next to FRAME. This will add a border round the edge of your map. The contents of your PRINT COMPOSER window should now look like this:

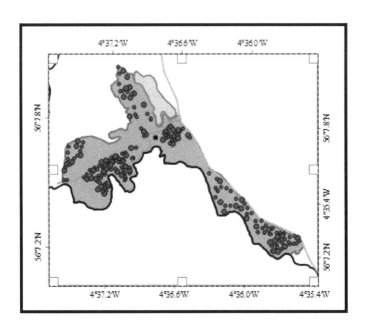

Next, you will add a scale bar. To do this, click on the LAYOUT menu on the main menu bar and select ADD SCALEBAR. Now, holding down the left hand mouse button, draw a box any where on your map to add the scale bar. At this stage, do not worry about its size or position, you will sort these out next. To do this, go back to the ITEM PROPERTIES tab, where you will notice that different options are now available. This is because the scale bar is now selected on the map (it will have an outline round it with white boxes at its corners). If you wanted to get the options for the map back, you would simply click on the map in the PRINT COMPOSER window to select it. However, for the moment, leave the scale bar selected so you can change its settings.

In the ITEM PROPERTIES tab for the scale bar, for STYLE select LINE TICKS UP. Now, scroll down to the SEGMENTS section of the tab. Here, change the options for SEGMENTS to LEFT 0 and RIGHT 1. For SIZE, enter the value 250, and for HEIGHT, enter a value of 3mm. These settings will mean that your scale bar will represent one block of 250m on your map. Next, scroll down to POSITION AND SIZE and click on the arrow next to it to activate the position and size options. For PAGE, enter 1 then for X enter 35, and for Y enter 184. The scale bar will now move to the bottom left hand corner of your map. Finally, scroll down and make sure that the box next to BACKGROUND is not selected, then click anywhere on your map other than on the scale bar.

The contents of your PRINT COMPOSITION window should now look like this:

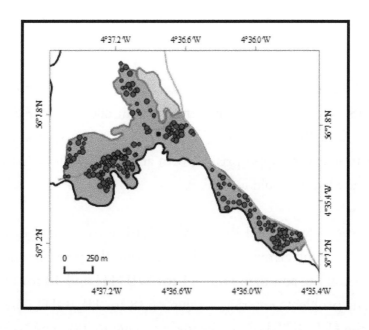

Exercise One: Creating A High Quality Map Of Your Data In QGIS

Once you have successfully inserted a scale bar, you will want to insert a North arrow to indicate which direction is north on your map. To do this, click on LAYOUT on the main menu bar and select ADD IMAGE. Now, holding down the left hand mouse button, draw a box anywhere on your map to add a blank North arrow. At this stage, do not worry about its size or position, or the fact that you cannot see the arrow, you will sort these out next. To do this, go back to the ITEM PROPERTIES tab, where you will notice that different options are now available. This is because the image you have just added is now selected on the map (it will have an outline round it with white boxes at its corners).

In the ITEM PROPERTIES tab, click on the button at the right hand end of the IMAGE SOURCE section (it has three dots in a row on it). This will open the SELECT SVG OR IMAGE FILE window. In this window, browse to C:/PROGRAM FILES/QGIS WIEN/APPS/QGIS/SVG/ARROWS (or C:/PROGRAM FILES/QGIS WIEN/APPS/QGIS-LTR/SVG/ARROWS, depending on how your computer is set up) and select NORTHARROW_04.SVG before clicking OPEN. (**NOTE:** If you are using a version of QGIS other than the version 2.8.3 recommended for this book, your QGIS folder will not be called QGIS WIEN, but will instead have the name of your version of QGIS, and you will have to navigate to that folder instead.) You will now be able to see a North arrow (or, depending on the size of the box you drew on your map, part of a North arrow) on your map. Next, for RESIZE MODE, select ZOOM, and for PLACEMENT, select MIDDLE.

NOTE: If you find that you cannot access the required SVG folder directly (for example, if you are using QGIS on a Mac OS computer), you can click on SEARCH DIRECTORY below the IMAGE SOURCE section. This will allow you to view the available image options, and you can select the appropriate North arrow image directly (see the map on the next page to find out what it should look like).

Once you have successfully selected the correct North arrow, scroll down to POSITION AND SIZE and click on the arrow next to it to activate the position and size options. For X enter 165, for Y enter 78, and then enter 11 for WIDTH and 18 for HEIGHT (**NOTE:** If you encounter any problems entering a value of 18 for HEIGHT, use the arrows at the right hand end of the HEIGHT box to change the value to 18). The North arrow will now have moved to the top right hand corner of your map. Finally, scroll down and make sure that the box next to BACKGROUND is not selected.

The contents of your PRINT COMPOSER window should now look like this:

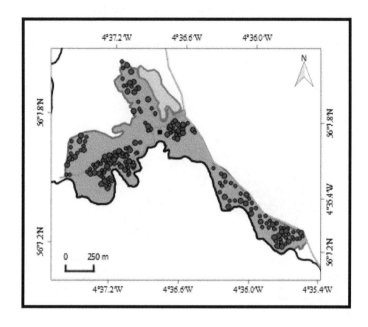

Since this is a relatively simple map, you can describe its contents in a figure legend which will tell the viewer what the different symbols and colours mean. For more complex maps, you might want to add a specific legend. This can be done through the LAYOUT menu.

All that is left now is for you to save your map and then export it. To do this, click on COMPOSER on the main menu bar and select SAVE PROJECT. Then click on COMPOSER again and select EXPORT AS IMAGE. This will open a window where you can select the format and the location where you wish to save it. The format you select will depend on what you wish to use your map for. For example, you may choose a different format depending on whether you are creating a map to include in a presentation or for a written report. For this exercise, export your map as a .jpg, call it BLUE_TIT_ OCCUPANCY_MAP and save it in the folder C:\QGIS_R_WORKBOOK. Once you have successfully completed this step, close the PRINT COMPOSER window. Finally, click on the PROJECT menu on the main menu bar on the main QGIS user interface and select SAVE to save the changes you have made to your GIS project.

Exercise One: Creating A High Quality Map Of Your Data In QGIS

Optional Extra:

If you wish to get more experience with creating high quality maps in QGIS, you can repeat this exercise, but using great tits (GT) as the target species. At the end of this process, your final map of great tit nest box occupancy should look like this:

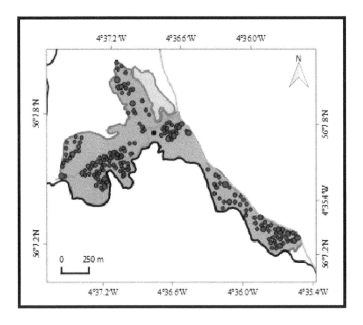

When you are completing this optional extra, remember to use different file names so that you do not over-write the data layer containing the information on blue tit nest box occupancy as you will need it for later exercises in this workbook.

--- Chapter Four ---

Exercise Two: Creating Raster Data Layers Of Environmental Variables In QGIS

In order to relate nest box occupancy and breeding success to local environmental variables, you first need to add environmental data layers to your GIS project and process them in a way that allows you to extract the required information from them. This is usually done by creating raster data layers to represent the required environmental variables, and this will be done in exercise two. A total of five environmental raster data layers will be created. These are raster data layers of local land elevation, the slope of the local terrain, hillshade (which measures how much direct sunlight each part of the study area will get at a specific time of day and year), the distance to the shoreline of Loch Lomond, a large body of water nearby, and the distance to the edge of the patch of native oak woodland in which the nest boxes are located.

When you are creating such environmental raster data layers, you need to consider the likely resolution of the relationships between them and the distribution of your target species. This is because this will define the cell size you will need to use when creating these raster data layers. If you select the wrong resolution, you may find that you cannot detect relationships which exist between them and the biology of your target species. Here, you will use a cell size of 10m by 10m. This has been selected because of the size of the oak woodland study area, and also because the aim of the analyses which you will conduct in the exercises three to five of this workbook is to look for potential relationships between the fine-scale local environmental variables and breeding behaviour. Thus, a 10m by 10m cell size is appropriate for this aim. If, for your own studies, you do not have an idea of what resolution would be most appropriate to use, you can derive this information empirically, through trial and error, by creating raster data layers at different resolutions and

Exercise Two: Creating Raster Data Layers Of Environmental Variables In QGIS

running exploratory analyses based upon them until you identify one that has the strongest relationship with your response variable(s).

The starting point for this exercise is the GIS project created in exercise one. To open this GIS project, first open QGIS. Once it has started, click on the PROJECT menu and select OPEN. When the OPEN window appears, browse to the location where your project is saved (C:\QGIS_R_WORKBOOK), select it and click OPEN. After you have opened the GIS project called EXERCISE_ONE, the first thing you need to do is save it under a new name. This is because you do not want to alter the contents of the original project, you just want to base your new one on it since this saves you having to add all the data layers again, and also having to reset the projection/coordinate system of the data frame. To save the project under a new name, click on PROJECT on the main menu bar and select SAVE AS. For this exercise, save it as EXERCISE_TWO in the C:\QGIS_R_WORKBOOK folder.

During this exercise, you will use three feature data layers to create your environmental raster data layers. These are a line data layer representing the shoreline of Loch Lomond, a polygon data layer representing the extent of the patch of native oak woodland where the nest boxes are located, and a line data layer representing fine-scale elevation contours. Two of these data layers were added to your GIS project in exercise one, and you will add the last one to it as part of this exercise. All of these data layers are already in the British National grid projection/coordinate system being used for the exercises in this book.

NOTE: It is important that you save your GIS project after you complete each step. This can be done by going to the PROJECT menu on the main menu bar and selecting SAVE.

STEP 1: ADD A LINE DATA LAYER OF LAND ELEVATION TO YOUR GIS PROJECT:

To start this exercise, you first need to add information about the local land elevation to your GIS project. This done by adding a data layer called SCENE_ELEVATION to it. This is a line data layer of elevation contours, and it can be added to your project by working through the flow diagram that starts at the top of the next page.

Exercise Two: Creating Raster Data Layers Of Environmental Variables In QGIS

Add a line data layer of land elevation to your GIS project

In this exercise, the elevation data layer which you wish to add to your GIS project is called SCENE_ELEVATION.SHP.

1. Open the ADD VECTOR LAYER window

On the main menu bar, click on LAYER and select ADD LAYER> ADD VECTOR LAYER. In the ADD VECTOR LAYER window that opens, browse to the location of your data layer (C:\QGIS_R_WORKBOOK) and then click on the section in the bottom right hand corner of the OPEN window and select ESRI SHAPEFILES (*.shp, .SHP). Now, select the data layer called SCENE_ELEVATION.SHP. Next, click OPEN in the browse window and then OPEN in the ADD VECTOR LAYER window.

2. Check the projection/coordinate system for the newly added data layer

Whenever you add a data layer to a GIS project, you should always check that it has a projection/coordinate system assigned to it, and look at what this projection/coordinate system is. This is so that you know whether you will need to assign a projection/coordinate system to it, or transform it into a different projection/coordinate system before you can use it in your GIS project. To check the projection/coordinate system of your newly added data layer, right click on its name in the TABLE OF CONTENTS window and select PROPERTIES. In the LAYER PROPERTIES window which opens, click on the GENERAL tab on the left hand side and check that there is a projection/coordinate system listed in the COORDINATE REFERENCE SYSTEM window. For the SCENE_ELEVATION data layer this should say :

EPSG:27700 – OSGB 1936 / British National Grid

This tells you that this data layer is in the British National Grid projection based on the OSGB 1936 datum. Now click OK to close the LAYER PROPERTIES window.

37

Exercise Two: Creating Raster Data Layers Of Environmental Variables In QGIS

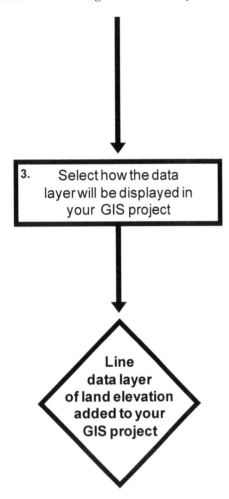

Right click on the name of the SCENE_ELEVATION data layer in the TABLE OF CONTENTS window, and select PROPERTIES. In the LAYER PROPERTIES window, click on the STYLE tab on the left hand side. In the top left hand corner of the STYLE tab, select CATEGORIZED from the drop down menu in the section which currently says SINGLE SYMBOL. For COLUMN, select ELEVATION from the drop down menu. Now click on the ADD button. This will add a line symbol in the section directly above this. In this section, double-click beside this line under VALUE and enter the number 0. Double-click under LEGEND and enter 0 here too. This tells QGIS to display the 0 elevation contour using this symbol. Next, click ADD again to add another line. Next to this, enter 20 for VALUE and 20 for LEGEND. Repeat this for 50, 100, 200 and 500. This will mean that only contours with these elevations will be displayed in the MAP window. Finally, select YlOrBr for the COLOR RAMP and then click OK to close the LAYER PROPERTIES window.

At the end of this step, your TABLE OF CONTENTS window should look like this:

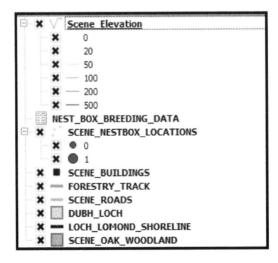

While the contents of your MAP window should look like this:

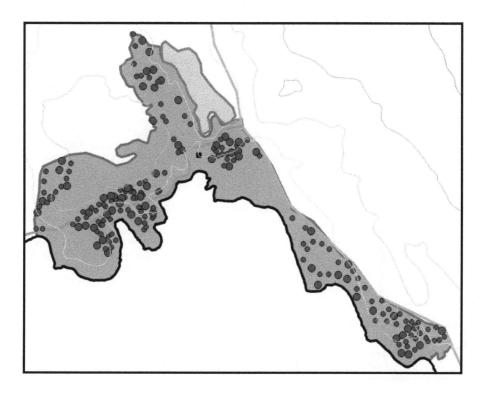

STEP 2: INTERPOLATE A RASTER DATA LAYER OF ELEVATION FROM YOUR LAND ELEVATION LINE DATA LAYER:

In this step, you will make your first raster data layer of land elevation in QGIS using the R.SURF.CONTOUR tool. For this exercise, a cell size of 10m by 10m will be used for the raster data layer's resolution. The extent will range from 696,700m in the north to 694,700m in the south and from 236,700m in the west to 239,500m in the east. This will ensure that it will cover the entire oak woodland study area. In the raster data layer created in this step, each cell within the given extent will have a value for land elevation, regardless of whether it is in on land or in Loch Lomond, a large lake that borders the oak woodland study area, based on its position relative to the elevation contours in the original contour data layer. This is okay, and you will remove the areas that fall in Loch Lomond from your elevation raster data layer in step 3. To create your first raster data layer of land elevation, work through the flow diagram that starts at the top of the next page.

Exercise Two: Creating Raster Data Layers Of Environmental Variables In QGIS

Land elevation contour data layer

For this exercise, the land elevation contour data layer is called SCENE_ELEVATION.

Before you can interpolate an elevation raster data layer from contour data using the R.SURF.CONTOUR tool, you first need to convert your contour data into a raster data layer. To do this, in the TOOLBOX window, select GRASS COMMANDS> VECTOR (V.*)> V.TO.RAST.ATTRIBUTE. In the V.TO.RAST.ATTRIBUTE window which opens, select SCENE_ELEVATION as the INPUT VECTOR LAYER. For NAME OF COLUMN FOR 'ATT' PARAMETER, select ELEVATION. For GRASS REGION EXTENT, enter the following text:

236700,239500,694700,696700

NOTE: There must be no spaces anywhere in this text otherwise this tool will not work. Next, enter 10 for GRASS REGION CELLSIZE. This will result in the creation of a raster data layer with a cell size of 10m by 10m. For RASTERIZED LAYER leave the default setting of [SAVE TO TEMPORARY FILE]. Now click on RUN to run the tool. This will create a temporary raster data layer where each cell has a value based on the contour it contains. If a cell does not contain a contour, it will be classified as 'no data' and will appear blank. This is okay.

1. Interpolate a depth raster data layer using the R.SURF.CONTOUR tool

Now that you have converted your contours into a raster data layer, you are ready to convert them into a continuous surface that has a value for land elevation for every grid cell in the raster data layer. In the TOOLBOX window, select GRASS COMMANDS> RASTER (R.*)> R.SURF.CONTOUR. In the R.SURF.CONTOUR window which opens, select the data layer you just created (RASTERIZED LAYER) as the RASTER LAYER WITH RASTERIZED CONTOURS. For GRASS REGION EXTENT, enter the following text:

236700,239500,694700,696700

NOTE: There must be no spaces anywhere in this text otherwise this tool will not work. Next, enter 10 for GRASS REGION CELLSIZE. This will result in the creation of a raster data layer with a cell size of 10m by 10m. For OUTPUT RASTER LAYER, type C:/QGIS_R_WORKBOOK/ELEVATION. Now click on RUN to run the tool. After it has finished running, right click on the data layer which this tool creates (called OUTPUT RASTER LAYER) in the TABLE OF CONTENTS window and select RENAME. Once its existing name has been selected, type ELEVATION so that it replaces it and then press the ENTER key on your keyboard. Finally, right click on the temporary raster data layer named RASTERIZED LAYER in the TABLE OF CONTENTS window and select REMOVE.

Land elevation raster data layer created from contour data layer

You can now compare your elevation raster data layer to the contour data layer (SCENE_ELEVATION) used to make it to check that the two are consistent with each other (the fact that you can do this quickly and easily is one of the main advantages of doing this step in QGIS rather than in R). This can be done by clicking on the name of the contour data layer in the TABLE OF CONTENTS window and dragging it until it is above the elevation raster data layer. Your TABLE OF CONTENTS window should look like this:

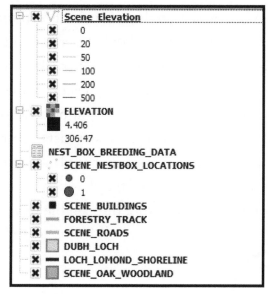

While the contents of your MAP window should look like this:

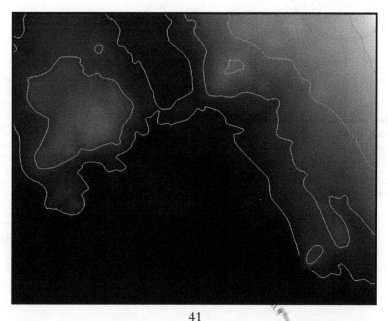

Exercise Two: Creating Raster Data Layers Of Environmental Variables In QGIS

STEP 3: CREATE A MASK AND USE IT TO REMOVE ANY AREAS WHERE IT IS INAPPROPRIATE TO INTERPOLATE LAND ELEVATION:

When you create a raster data layer using the R.SURF.CONTOUR tool, you will notice that it creates an interpolated raster data layer that covers the full extent you entered when using this tool. However, this is not always appropriate. For example, in this exercise, it has resulted in elevation data being interpolated for the area covered by Loch Lomond where no land elevation contours were present in the original data set. Interpolations in such areas can create artefacts in your data, and so they should be removed from your elevation raster data layer. This is done by using something called a mask. A mask basically acts like a 'cookie-cutter', allowing you to cut out the bits of a raster data layer you want to keep and changing all other cells to a no data value. These might be areas of land from a raster layer of water depth, areas of water from a land raster data layer or simply areas of an environmental raster data layer that fall outside your study area. In this step, you will create a mask that you can use to remove the areas of the elevation raster data layer created in step two that fall in Loch Lomond. To do this, work through the flow diagram provided below.

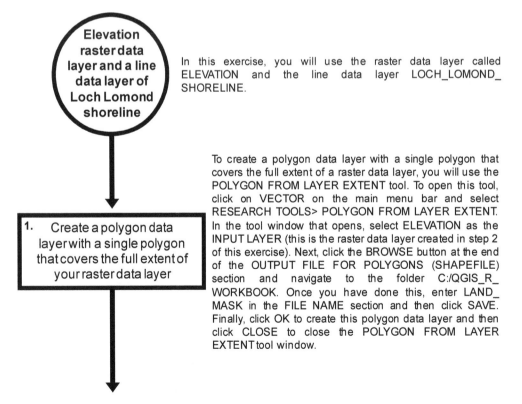

2. Remove the parts of the polygon which fall in Loch Lomond

You now want to remove the parts of your new polygon that fall in Loch Lomond. To do this, you first need to load and activate the DIGITIZING TOOLS plugin. To do this, click on PLUGINS on the main menu bar and select MANAGE AND INSTALL PLUGINS. In the PLUGINS window that opens, enter the term DIGITIZING TOOLS in the search section and then click on DIGITIZING TOOLS when it appears in the section below this. If it is not already installed, click on INSTALL PLUGIN to install it. Make sure that there is a cross next to its name (which indicates it has been activated) and then click on CLOSE to close the PLUGINS window. The DIGITIZING TOOLS toolbar should be added automatically to the OPTIONAL TOOLBARS area of the QGIS user interface. If it isn't, you can activate it by clicking on VIEW on the main menu bar and selecting TOOLBARS> DIGITIZING TOOLS.

You are now ready to remove the parts of your new polygon that fall in Loch Lomond. To do this, right click on the name of the data layer LAND_MASK in the TABLE OF CONTENTS window and select TOGGLE EDITING. Now click on VIEW on the main menu bar and then SELECT> SELECT FEATURE(S). Once the selection tool has been activated, click on the polygon created in stage 1 to select it. Now click on VECTOR on the main menu bar and select SPATIAL QUERY> SPATIAL QUERY. Once the SPATIAL QUERY tool window opens, select LOCH_LOMOND_ SHORELINE for SELECT SOURCE FEATURES FROM, and then for WHERE THE FEATURE select INTERSETCS. Finally, select LAND_MASK for REFERENCE FEATUERS OF. This will select all the lines in the LOCH_ LOMOND_SHORELINE data layer that intersect with your LAND_MASK polygon.

Next, find where the DIGITIZING TOOLS toolbar has been added to the main QGIS USER INTERFACE (this should be in the OPTIONAL TOOLBARS AREA – see figure 2 on page 7). On this toolbar, find the SPLIT SELECTED FEATURES WITH SELECTED LINE FROM ANOTHER LAYER tool and click on it. In the window that opens, select LOCH_ LOMOND_SHORELINE and then click OK. This will split the polygon you created in stage 1 of this step into four parts.

To remove the parts that fall in Loch Lomond, right click on the name of the data layer LAND_MASK in the TABLE OF CONTENTS window and select OPEN ATTRIBUTE TABLE. In the TABLE window that opens, select the rows that start with 1, 2 and 3 (but not the one that starts 0) and then close the TABLE window. Next, click on EDIT on the main menu bar and select DELETE SELECTED. This will leave a single polygon in the LAND_MASK data layer that only covers the land and none of Loch Lomond. Right click on the name of this data layer in the TABLE OF CONTENTS window and select TOGGLE EDITING MODE. When asked if you wish to save your edits, click SAVE.

Exercise Two: Creating Raster Data Layers Of Environmental Variables In QGIS

3. Create a mask to remove areas of the interpolated elevation raster data layer which fall in Loch Lomond

To remove any areas of your interpolated elevation raster which fall in Loch Lomond, you first need to create a mask raster data layer from your polygon data layer. To do this, click on RASTER on the main menu bar and select CONVERSION> RASTERIZE (VECTOR TO RASTER). In the tool window which opens, select LAND_MASK from the drop down menu in the INPUT FILE (SHAPEFILE) section, and then select AREA in the ATTRIBUTE section. Now, in the OUTPUT FILE FOR RASTERIZED VECTORS (RASTER) section, enter C:/QGIS_R_WORKBOOK/ LAND_MASK_RASTER. Next, click on the button beside RASTER RESOLUTION IN MAP UNITS PER PIXEL and then enter 10 for HORIZONTAL and 10 for VERTICAL. Once you have done this, you need to tell it to mark any cell with a zero in it as no data, and also set it to create a raster with a specific extent. This is done by clicking on the EDIT button (it is towards the lower right hand corner of the tool window and has a picture of a yellow pencil on it). Once you are in editing mode, enter a space and then add the following code to the lowest section of the tool window (after all the code which is currently there):

```
-a nodata 0 -te 236700 694700 239500 696700
                -burn 1
```

Once you have added this code, you can click on the OK button to run the tool. **NOTE:** The values after the expression `-te` are the minimum X value, the minimum Y value, the maximum X value and the maximum Y value for the extent that you want your raster data layer to have. **NOTE:** The term `-burn 1` sets all values in the resulting raster data layer that are not classified as 'No Data' to 1. This is the value required to allow the raster data layer being created at this stage to be used in stage 4 to mask the elevation raster data layer created in step 2 (see page 40). Finally, click CLOSE on the RASTERIZE (VECTOR TO RASTER) window.

Exercise Two: Creating Raster Data Layers Of Environmental Variables In QGIS

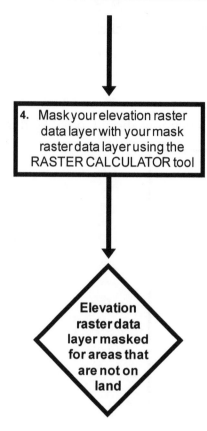

Now you have created a suitable mask, you can use the RASTER CALCULATOR tool to remove the areas of your elevation raster data layer which fall in Loch Lomond. To do this, click on RASTER on the main menu bar and select RASTER CALCULATOR. In the RASTER CALCULATOR window which will open, double-click on "ELEVATION@1" in the top left hand section so that it is added to the lower expression section. Next click on the * button in the middle of the window and then double-click on "LAND_MASK_RASTER@1" in the top left section to add it to the expression. The final expression should look like this:

"ELEVATION@1" * "LAND_MASK_RASTER@1"

Next, in the top right hand section of the tool window, type C:/QGIS_R_WORKBOOK/ELEVATION_MASKED in the OUTPUT LAYER box, then click on the OK button to run the tool.

Once you have completed this step, remove the data layer called LAND_MASK from your GIS project by right-clicking on its name in the TABLE OF CONTENTS window and selecting REMOVE. Now, right click on ELEVATION_MASKED data layer and select PROPERTIES. When the LAYER PROPERTIES window opens, click on the STYLE tab. In the STYLE tab, select SINGLEBAND PUSEDOCOLOR from the drop down menu beside RENDER TYPE. Next, directly under GENERATE NEW COLOR MAP on the right hand side, select the colour ramp called YlOrBr (you will need to scroll down to find it) and then click CLASSIFY. Finally, click OK to close the LAYER PROPERTIES window. Click on the name of your ELEVATION_MASKED data layer in the TABLE OF CONTENTS window and drag it to the top of the list of data layers in your GIS project. Now do the same for the LOCH_LOMOND_SHORELINE data layer. You should now be able to see where the cells that fall on in Loch Lomond have been removed from the ELEVATION_MASKED raster data layer, as you will be able to see the cells of the ELEVATION raster data layer below them. You can now remove the raster data layer called ELEVATION from your GIS project by right-clicking on its name in the TABLE

Exercise Two: Creating Raster Data Layers Of Environmental Variables In QGIS

OF CONTENTS window and selecting REMOVE. Finally, right click on ELEVATION_MASKED and select ZOOM TO LAYER before saving the changes you have made to your GIS project.

At the end of this step, your TABLE OF CONTENTS window should look like this:

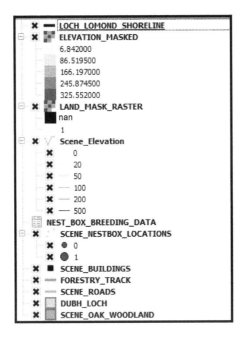

The contents of your MAP window should look like this:

STEP 4: DERIVE A RASTER DATA LAYER OF SLOPE FROM YOUR ELEVATION RASTER DATA LAYER:

The slope is calculated as the maximum difference in elevation between a grid cell and those that immediately surround it. This can be expressed either as the absolute change in elevation over a specific distance (e.g. 10 metres per kilometre), as a percentage (e.g. 1% change in elevation per kilometre) or as the angle from a horizontal straight line that would connect the two points (e.g. 5 degrees). In theory, deriving a data layer of slope from a raster of land elevation (or water depth) is very easy. This is because many GIS software packages, including QGIS, have a specific tool for generating raster data layers of slope. However, in practice, things can be a little more complicated. Firstly, in order to ensure that only suitable data are used to calculate the slope values, it is important to ensure that your elevation data layer has been masked to remove any unsuitable data from it before you create a slope raster data layer (these can be areas of water for a land elevation raster data layer or areas of land for a water depth raster data layer). Secondly, you need to ensure that your elevation raster data layer has been created in a projection/coordinate system that has real units of distance for its map units (e.g. metres) and not one that uses decimal degrees. This is because slope is calculated by comparing the difference in elevation to the distance between adjacent cells and if this is not measured in a real unit of distance, the values obtained will not be correct.

For this exercise, your elevation raster data layer is already in an appropriate projection/coordinate system (for this exercise, this is the British National Grid projection/coordinate system), and has already been masked to remove unsuitable data (see step 3 of this exercise)

NOTE: Before you start this step, you will need to ensure that the TERRAIN ANALYSIS plugin is loaded into QGIS. To do this, click on PLUGINS on the main menu bar and select MANAGE AND INSTALL PLUGINS. In the window that opens, type TERRAIN into the search section, and then select the plugin called RASTER TERRAIN ANALYSIS PLUGIN. If this is not already installed, click on the INSTALL PLUGIN button. If it is already installed, make sure that is it active (i.e. that it has a cross or a tick next to its name). You can now click CLOSE to close the PLUGINS window. If you cannot access this plugin, type SLOPE into the search box at the top of the TOOLBOX window. This will

Exercise Two: Creating Raster Data Layers Of Environmental Variables In QGIS

provide you with access to a range of additional tools for generating a raster data layer of slope. You can now generate your raster data layer of slope by working through the following flow diagram:

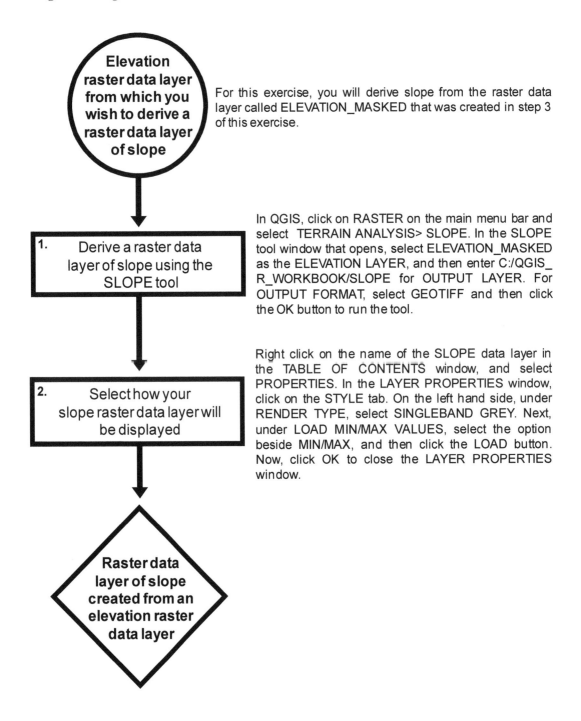

When you have completed this step, make sure that the data layer called SLOPE is at the top of the list of data layers in your TABLE OF CONTENTS window. It should now look like this:

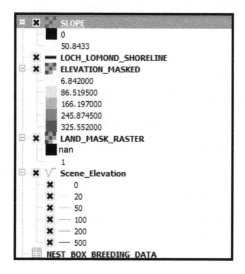

While the contents of your MAP window should look like this:

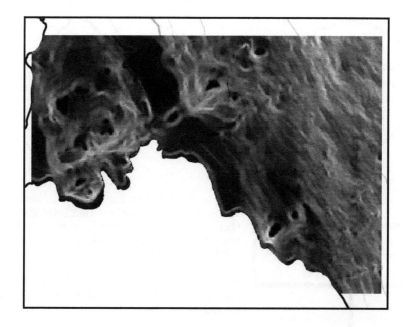

STEP 5: CREATING A RASTER DATA LAYER OF HILLSHADE FROM AN ELEVATION DATA LAYER:

Hillshade uses information about the surrounding land elevation to work out how much direct sunlight each grid cell within a raster data layer will receive when the sun is in a specific position in the sky. The position of the sun is be determined by two components: Time of day, which provides a measure of how high the sun is above the horizon, and time of year, which provides information about the direction the sun will be shining from at that time of day. Together, these give the altitude and azimuth of the sun which can be used to derive a hillshade raster data layer from a land elevation raster data layer. For this exercise, you will use an altitude of 60 degrees and an azimuth of 180 degrees. This will provide a measure of how much sunlight each grid cell within the native oak woodland study area gets at around midday in early summer. This can be done by working through the flow diagram provided below.

NOTE: If you wish to work out what altitude and azimuth to use to calculate hillshade for any specific location, date and time of day, you can use the information on this website to help you: *www.suncalc.org/#/40.1789,-3.5156,3/2019.08.16/11:00/1/3*.

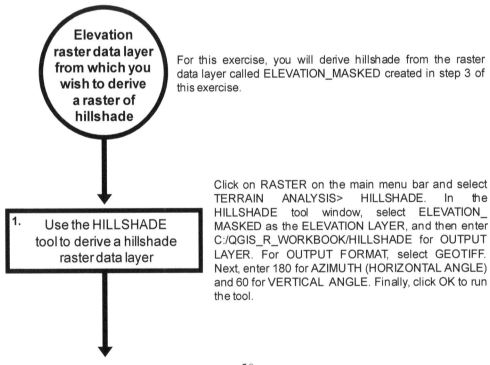

Exercise Two: Creating Raster Data Layers Of Environmental Variables In QGIS

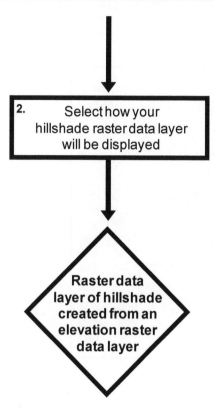

Right click on the name of the HILLSHADE data layer in the TABLE OF CONTENTS window, and select PROPERTIES. In the LAYER PROPERTIES window, click on the STYLE tab on the left hand side. In the top left hand corner of the STYLE tab, select SINGLEBAND GRAY from the drop down menu next to RENDER TYPE. For COLOR GRADIENT, select BLACK TO WHITE and for CONTRAST ENHANCEMENT, select STRETCH TO MINMAX. Finally, click OK to close the LAYER PROPERTIES window.

Once you have completed this step, re-arrange the order of the data layers in the TABLE OF CONTENTS window until it looks like this:

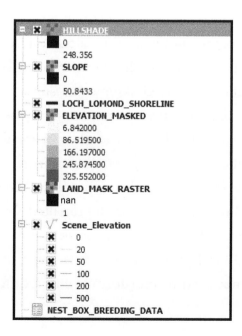

Exercise Two: Creating Raster Data Layers Of Environmental Variables In QGIS

The contents of your MAP window should now look like this:

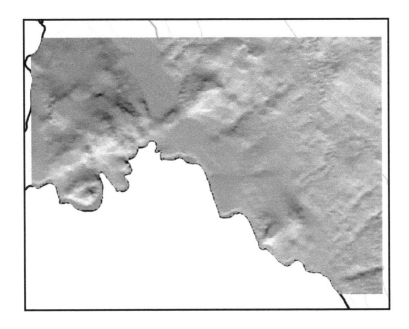

STEP 6: CREATING A RASTER DATA LAYER OF DISTANCE TO THE EDGE OF LOCH LOMOND:

In this step, you will create a raster data layer that will measure the distance to the edge of Loch Lomond. The same process can be used to calculate raster data layers of distances to any type of environmental feature. These distances are calculated as Euclidean distances. This means that the calculations involved assume the surface on which the distances are calculated is flat. As a result, distance values can only be calculated in projection/coordinate systems that use real world units as their map units and not ones that use decimal degrees. In addition, the projection/coordinate system must not distort measurements of straight-line distances. In this step, you will create a raster data layer of the distance to the edge of Loch Lomond as represented by the LOCH_LOMOND_SHORELINE line data layer. This is already in an appropriate projection/coordinate system (the British National Grid projection which is being used for the exercises in this book). This means you do not need to transform it into a different projection/coordinate

Exercise Two: Creating Raster Data Layers Of Environmental Variables In QGIS

system before you can create your raster data layer of the distance to the edge of Loch Lomond. To create this raster data layer, work through the following flow diagram:

Data layer representing the shoreline of Loch Lomond

For this exercise, the data layer you will use is called LOCH_LOMOND_SHORELINE.

Before you can create a raster data layer of distance to the edge of the loch, you first need to create a raster data layer of the loch's shoreline. To do this, click on RASTER on the main menu bar and select CONVERSION> RASTERIZE (VECTOR TO RASTER). In the tool window which opens, select LOCH_LOMOND_SHORELINE from the drop down menu in the INPUT FILE (SHAPEFILE) section. Now, in the OUTPUT FILE FOR RASTERIZED VECTORS (RASTER) section, enter C:/QGIS_R_WORKBOOK/LOCH_EDGE_RASTER. Next, click on the button beside RASTER RESOLUTION IN MAP UNITS PER PIXEL and then enter 10 for HORIZONTAL and 10 for VERTICAL. After you have done this, you need to set the tool to create a raster with a specific extent. This is done by clicking on the EDIT button (it is towards the lower right hand corner of the tool window and has a picture of a yellow pencil on it). Once you are in editing mode, enter a space and then add the following code to the lowest section of the tool window (after all the code which is currently there):

```
-a_nodata 0 -te 236700 694700 239500 696700
                -burn 1
```

1. Create a raster data layer of distance to the edge of Loch Lomond

Once you have added this code, you can click on the OK button to run the tool. **NOTE:** The values after the expression `-te` are the minimum X value, the minimum Y value, the maximum X value and the maximum Y value for the extent that you want your raster data layer to have. **NOTE:** The term `-burn 1` sets the values for all the grid cells in the resulting raster data layer that contain part of the shoreline to 1. Now click CLOSE on the RASTERIZE (VECTOR TO RASTER) window.

Next, in the TOOLBOX window, double-click on GRASS COMMANDS> RASTER (R.*)> R.GROW.DISTANCE. In the tool window that opens, select LOCH_EDGE_RASTER as the INPUT RASTER LAYER and EUCLIDEAN as the METRIC. Leave the GRASS REGION EXTENT with its default setting, but for GRASS REGION CELLSIZE enter 10. Finally, in the DISTANCE LAYER section, type select SAVE TO TEMPORAY FILE and then click on the RUN button. Once this tool has finished running, right click on the data layer called OUTPUT LAYER in the TABLE OF CONTENTS window and select REMOVE to remove it from your GIS project.

Exercise Two: Creating Raster Data Layers Of Environmental Variables In QGIS

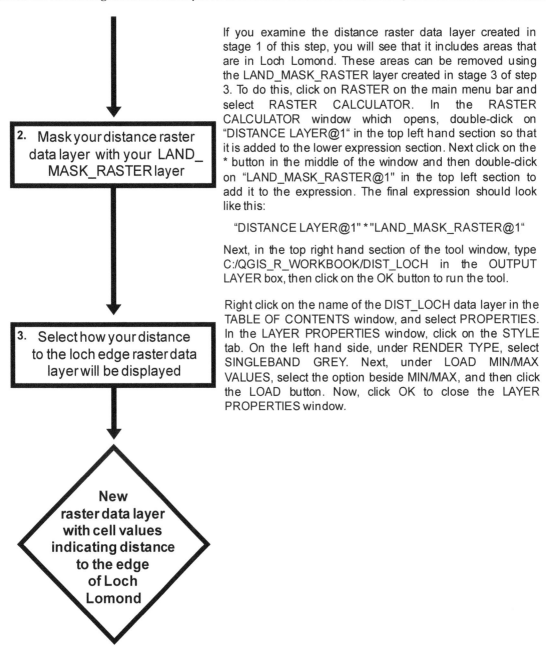

2. Mask your distance raster data layer with your LAND_MASK_RASTER layer

If you examine the distance raster data layer created in stage 1 of this step, you will see that it includes areas that are in Loch Lomond. These areas can be removed using the LAND_MASK_RASTER layer created in stage 3 of step 3. To do this, click on RASTER on the main menu bar and select RASTER CALCULATOR. In the RASTER CALCULATOR window which opens, double-click on "DISTANCE LAYER@1" in the top left hand section so that it is added to the lower expression section. Next click on the * button in the middle of the window and then double-click on "LAND_MASK_RASTER@1" in the top left section to add it to the expression. The final expression should look like this:

"DISTANCE LAYER@1" * "LAND_MASK_RASTER@1"

Next, in the top right hand section of the tool window, type C:/QGIS_R_WORKBOOK/DIST_LOCH in the OUTPUT LAYER box, then click on the OK button to run the tool.

3. Select how your distance to the loch edge raster data layer will be displayed

Right click on the name of the DIST_LOCH data layer in the TABLE OF CONTENTS window, and select PROPERTIES. In the LAYER PROPERTIES window, click on the STYLE tab. On the left hand side, under RENDER TYPE, select SINGLEBAND GREY. Next, under LOAD MIN/MAX VALUES, select the option beside MIN/MAX, and then click the LOAD button. Now, click OK to close the LAYER PROPERTIES window.

New raster data layer with cell values indicating distance to the edge of Loch Lomond

NOTE: Do not worry if the contents of your MAP window appear to be completely black. This is okay. To correct this, right click on the name of the data layer called OUTPUT_LAYER in the TABLE OF CONTENTS window and select REMOVE. You should now be able to see the contents of the DIST_LOCH data layer which you have created in this step.

At the end of this step, right click on the data layer called DISTANCE LAYER in the TABLE OF CONTENTS window and select REMOVE. Finally, right click on the name of the data layer LOCH_EDGE_RASTER and select REMOVE to delete this data layer from your GIS project. Your TABLE OF CONTENTS window should look like this:

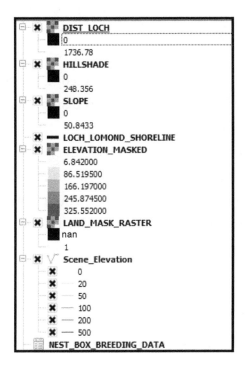

While the contents of your MAP window should look like this:

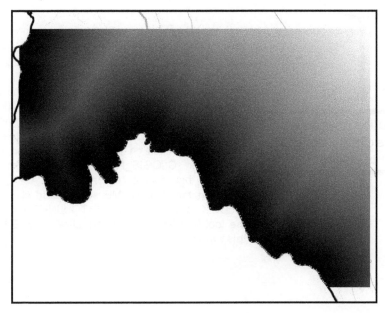

STEP 7: CREATE A RASTER DATA LAYER OF DISTANCE TO THE EDGE OF THE PATCH OF OAK WOODLAND CONTAINING THE NEST BOXES:

The last environmental data layer you will create is one that measures the distance to the edge of the patch of native oak woodland containing the nest boxes. As with the raster data layer of the distance to the edge of Loch Lomond, this raster data layer will be created in two stages. Firstly, a raster data layer will be created with the R.GROW.DISTANCE tool to calculate the distance that each grid cell is from the edge of the patch of woodland. When this data layer is made, you will see that it calculates distance values for all cells in the raster data layer and not just the ones which are inside of the area of woodland. In order to remove all the distance values from the cells outside the woodland area, in the second stage, you will create a mask that you can use to mask the distance raster data layer so that only the distance values inside the wood are retained. This results in a final raster data layer where the values represent the distance of each grid cell within the woodland to its nearest edge. To create this raster data layer, work through the following flow diagram:

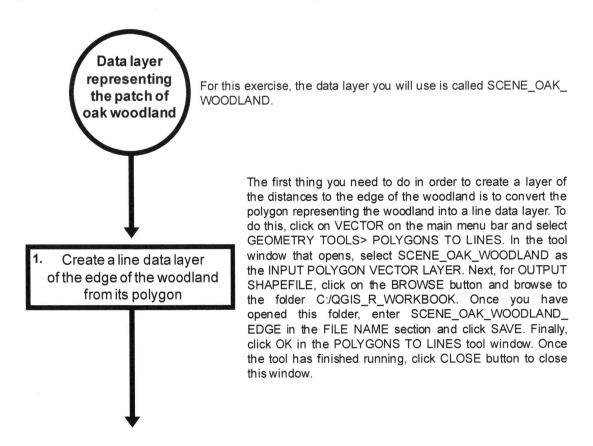

Exercise Two: Creating Raster Data Layers Of Environmental Variables In QGIS

2. Create a raster data layer of the distance to the edge of the woodland

Before you can create a raster data layer of distance to the edge of the woodland, you first need to create a raster data layer of the woodland edge itself. To do this, click on RASTER on the main menu bar and select CONVERSION> RASTERIZE (VECTOR TO RASTER). In the tool window which opens, select SCENE_OAK_WOODLAND_EDGE from the drop down menu in the INPUT FILE (SHAPEFILE) section. Now, in the OUTPUT FILE FOR RASTERIZED VECTORS (RASTER) section, enter C:/QGIS_R_ WORKBOOK/WOODLAND_EDGE_RASTER. Next, click on the button next to RASTER RESOLUTION IN MAP UNITS PER PIXEL and then enter 10 for HORIZONTAL and 10 for VERTICAL. After you have done this, you need to set the tool to create a raster with a specific extent. This is done by clicking on the EDIT button (it is towards the lower right hand corner of the tool window and has a picture of a yellow pencil on it). Once you are in editing mode, enter a space and then add the following code to the lowest section of the tool window (after all the code which is currently there):

```
-a_nodata 0 -te 236700 694700 239500 696700
-burn 1
```

Once you have added this code, you can click on the OK button to run the tool. **NOTE:** The values after the expression -te are the minimum X value, the minimum Y value, the maximum X value and the maximum Y value for the extent that you want your raster data layer to have. **NOTE:** The term -burn 1 sets the values for all the grid cells in the resulting raster data layer that contain part of the woodland's edge to 1. Now click CLOSE on the RASTERIZE (VECTOR TO RASTER) window.

Next, in the TOOLBOX window, double-click on GRASS COMMANDS> RASTER (R.*)> R.GROW.DISTANCE. In the tool window that opens, select WOODLAND_EDGE_ RASTER as the INPUT RASTER LAYER and EUCLIDEAN as the METRIC. Leave the GRASS REGION EXTENT with its default setting, but for GRASS REGION CELLSIZE enter 10. Finally, in the DISTANCE LAYER section, select SAVE TO TEMPORARY FILE and then click on the RUN button.

Exercise Two: Creating Raster Data Layers Of Environmental Variables In QGIS

3. Create a mask to remove areas of the interpolated distance raster which are outside the woodland

To remove any areas of your distance raster which are outside the woodland, you first need to create a mask raster data layer from the polygon data layer representing the woodland area. To do this, click on RASTER on the main menu bar and select CONVERSION> RASTERIZE (VECTOR TO RASTER). In the tool window which opens, select SCENE_OAK_WOODLAND from the drop down menu in the INPUT FILE (SHAPEFILE) section. Now, in the OUTPUT FILE FOR RASTERIZED VECTORS (RASTER) section, enter C:/QGIS_R_WORKBOOK/WOODLAND_MASK. Next, click on the button beside RASTER RESOLUTION IN MAP UNITS PER PIXEL and then enter 10 for HORIZONTAL and 10 for VERTICAL. After you have done this, you need to tell it to mark any cell with a zero in it as no data, and also set it to create a raster with a specific extent. This is done by clicking on the EDIT button (it is towards the lower right hand corner of the tool window and has a picture of a yellow pencil on it). Once you are in editing mode, enter a space and then add the following code to the lowest section of the tool window (after all the code which is currently there):

```
-a_nodata 0 -te 236700 694700 239500 696700
                -burn 1
```

Once you have added this code, you can click on the OK button to run the tool. **NOTE:** The values after the expression -te are the minimum X value, the minimum Y value, the maximum X value and the maximum Y value for the extent that you want your raster data layer to have. **NOTE:** The term -burn 1 sets the values for all the grid cells in the resulting raster data layer that contain part of the woodland polygon to 1. Finally, click CLOSE on the RASTERIZE (VECTOR TO RASTER) window.

Exercise Two: Creating Raster Data Layers Of Environmental Variables In QGIS

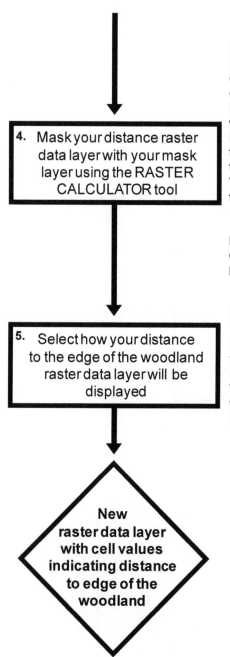

Now you have created a suitable mask, you can use the RASTER CALCULATOR tool to remove the areas from the distance layer that fall outside of the area of woodland. To do this, click on RASTER on the main menu bar and select RASTER CALCULATOR. In the RASTER CALCULATOR window which will open, double-click on "WOODLAND_MASK@1" in the top left hand section so that it is added to the lower expression section. Next click on the * button in the middle of the window and then double-click on "DISTANCE LAYER@1" in the top left section to add it to the expression. The final expression should look like this:

"WOODLAND_MASK@1" * "DISTANCE LAYER@1"

Next, in the top right hand section of the tool window, type C:/QGIS_R_WORKBOOK/DIST_EDGE in the OUTPUT LAYER box, then click on the OK button to run the tool.

Right click on the name of the DIST_EDGE data layer in the TABLE OF CONTENTS window, and select PROPERTIES. In the LAYER PROPERTIES window, click on the STYLE tab. On the left hand side, under RENDER TYPE, select SINGLEBAND GREY. Next, under LOAD MIN/MAX VALUES, select the option beside MIN/MAX, and then click the LOAD button. Now, click OK to close the LAYER PROPERTIES window.

At the end of this step, remove the data layers called LAND_MASK_RASTER, WOODLAND_MASK, DISTANCE LAYER, OUTPUT VALUE and WOODLAND_EDGE_RASTER from your GIS project. To do this, right click on their names in the TABLE OF CONTENTS window and select REMOVE.

Exercise Two: Creating Raster Data Layers Of Environmental Variables In QGIS

Your TABLE OF CONTENTS window should now look like this:

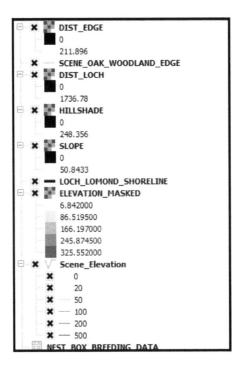

In the TABLE OF CONTENTS window, uncheck the boxes next to all the raster data layers of environmental variables created in this exercise except DIST_EGDE so that it is the only one which is displayed in the MAP window. It should now look like this:

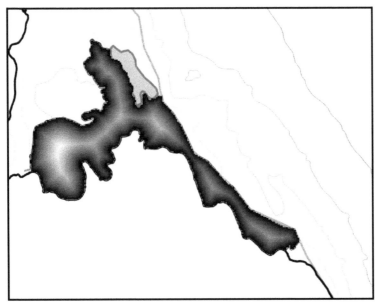

You can now remove the data layer called SCENE_OAK_WOODLAND_EDGE from your GIS project as it is no longer needed. This is done by right-clicking on its name in the TABLE OF CONTENTS window and selecting REMOVE. The final DIST_EDGE raster data layer can be used to investigate whether nest box occupancy and breeding success is related to the distance to the edge of the woodland in which they are sited

NOTE: Sometimes, the raster processing tools in QGIS can have problems transferring the projection/coordinate system information to the new raster data layers they create. As a result, once you have finished creating a set of raster data layers, you should check that all the new raster data layers generated have the correct projection/coordinate system assigned to them. To do this, right click on the name of a raster data layer in the TABLE OF CONTENTS window and select PROPERTIES. In the LAYER PROPERTIES window that open, click on the GENERAL tab and examine the contents of the COORDINATE REFERENCE SYSTEM section. It should say SELECTED CRS (EPSG: 27700 – OSGB 1936 / BRITSH NATIONAL GRID). This is because this is the projection/coordinate system being used for the current GIS project. If doesn't say this, you can assign the correct projection/coordinate system the raster data layer by clicking on this box and selecting PROJECT CRS (EPSG: 27700 – OSGB 1936 / BRITSH NATIONAL GRID) from the drop down menu that appears. Do this for all the environmental raster data layers generated in this exercise (ELEVATION_MASKED, SLOPE, HILLSHADE, DIST_ LOCH and DIST_EDGE).

Optional Extra:

If you wish to get more experience at creating raster data layers of environmental variables for use in an spatial analysis project, you can repeat the main exercise but using a 40m^2 resolution (that is, with a cell size of 20m by 20m). You can also make raster data layers with distances to other features in the study area, such as the distance to the nearest road and the distance to the forestry track which passes through the middle of the patch of native oak woodland where the nest boxes are located. These could be used to investigate whether the level of disturbance associated with these human landscape features impacts nest box occupancy and breeding success.

Exercise Two: Creating Raster Data Layers Of Environmental Variables In QGIS

NOTE: If you do this optional extra, remember to use different names for your data layers so that you do not over-write any of the data layers created in the main exercise. This is because these data layers will be needed for exercises three and four.

--- Chapter Five ---

Exercise Three: Linking Spatial Data Sets In QGIS And Conducting Basic Descriptive Analyses Using R

In exercise one, you created a data layer which provided information about the locations of nest boxes which you will use to investigate the relationship between breeding behaviour of hole-nesting birds and environmental variables in the remaining exercises in this book. In exercise two, you created raster data layers of local environmental variables that might influence which nest boxes are occupied by a particular species and its level of breeding success within each one. In this exercise, you will bring this information together to create a single 'big table' containing all the information you need to start analysing the relationships between environmental variables and breeding behaviour in these nest boxes. In this 'big table', each field will represent a separate variable, while each row will represent a different nest box within the oak woodland study area.

Once this 'big table' has been created you can start to analyse it. For this exercise, you will start by creating graphs of two measures of how nest box occupancy by blue tits is related to elevation. The first graph will show a simple frequency distribution of how many nest boxes were occupied by this species in different elevation ranges during one particular breeding season. This will give you a first look at the relationship between these two variables. However, it does not account for the number of nest boxes available at each elevation, and this can create biases in such simple frequency distributions. As a result, you will also create a second graph which shows the relationship between nest box occupancy and elevation. In this context, occupancy is the proportion of available boxes within an elevation category that were actually occupied by the target species (in this instance, blue tits).

Exercise Three: Linking Spatial Data Sets In QGIS And Conducting Basic Descriptive Analyses In R

These graphs will be created using R. While R is a command-driven software package which may seem daunting to use at first, it is actually quite straight forward to use if you know the code you need to make it do what you wish. For this exercise, you will be provided with all the required code both in this workbook (as part of the flow diagrams and in the appendix I) and in a document (called R_CODE_GIS_WORKBOOK_1.DOC) which can be found in the folder which you downloaded at the start of exercise one.

The starting point for this exercise is the GIS project created in exercise two. If you have not already done so, open QGIS. Once it is open, click on the PROJECT menu and select OPEN. When the OPEN window appears, browse to the location where your project is saved (C:\QGIS_R_WORKBOOK), select it and click OPEN. Once you have opened the GIS project called EXERCISE_TWO, the first thing you need to do is save it under a new name. This is because you do not want to alter the contents of the original project, you just want to base your new one on it since this saves you having to add all the data layers again, and also having to reset the projection/coordinate system of the data frame. To save the project under a new name, click on PROJECT on the main menu bar, and select SAVE AS. For this exercise, save it as EXERCISE_THREE in the C:\QGIS_R_WORKBOOK folder.

For this exercise, you will use the data layer showing the location of the nest boxes created in exercise one (SCENE_NESTBOX_LOCATIONS) and the elevation raster data layer with a 10m resolution created in the main part of exercise two (ELEVATION_MASKED).

For those steps which take place in the GIS environment, images of the contents of the MAP window, the TABLE OF CONTENTS window and/or the TABLE window will be provided so that you have an idea of what your GIS project should look like at that specific point as you progress through this exercise. Similarly, for those steps which require the use of R, images of the expected R outputs will be provided.

When it comes to entering the R code required to complete this, and later, exercises, you have two options. The first is to type them in yourself following the code provided in the instruction sets for each exercise. However, R is very picky in terms of having to get the text exactly right before it will work (this includes getting all the uppercase, lowercase,

Exercise Three: Linking Spatial Data Sets In QGIS And Conducting Basic Descriptive Analyses In R

underscores and spaces in the appropriate places). This means it is very easy to make mistakes when typing in the code. Therefore, as an alternative, you can copy the lines of code that make up each command from the document R_CODE_GIS_ WORKBOOK_1.DOC and paste them into your R Console, just as you would do in any word-processing document. This will make it much easier for you to concentrate on learning how to integrate QGIS and R as you will not have to spend as much time troubleshooting and correcting the code that you need to run each command.

In this document, the individual parts of the code have been colour-coded to help you see what each part does. In this colour-coding, red is always used for a file name, blue for the name of an object which has been created in the R workspace (such as a data set or a specific model), green for the R commands themselves, orange for names of any ecological variables (such as measures of species occurrence or environmental variables) referred to in within the commands and black for the settings used to run the commands.

STEP 1: LINK NEST BOX LOCATION DATA LAYER AND ELEVATION DATA LAYER TOGETHER TO CREATE A 'BIG TABLE' IN QGIS:

In this step, you will create the 'big table' you need to start analysing the spatial relationships within the data set being used in this workbook. This involves the use of the POINT SAMPLING TOOL tool. If you have created your raster data layers of environmental variables properly, so that they all have the same extent, compatible cell sizes and are all in the same projection/coordinate system, this is surprisingly quick and easy to do. However, this is where any incompatibilities between your data layers will become apparent and you may find you have to go back and repeat earlier steps in order to get it to work properly.

To link the nest box data to the elevation raster data layer, you first need to download and install the POINT SAMPLING TOOL plugin. To install this plugin, click on PLUGINS on the main menu bar and select MANAGE AND INSTALL PLUGINS. This will open the PLUGINS window. In this window, scroll down until you find the POINT SAMPLING TOOL. Select it and then click on the INSTALL PLUGIN button in the bottom right-hand corner. Once the plugin has been installed, click CLOSE to close the

Exercise Three: Linking Spatial Data Sets In QGIS And Conducting Basic Descriptive Analyses In R

PLUGINS window, and then make sure that the point data layer containing your SCENE_NESTBOX_LOCATIONS and ELEVATION_MASKED data layers are set to display (i.e. they have a cross in the box beside their names in the TABLE OF CONTENTS window). Next, click on PLUGINS on the main menu bar and select ANALYSES> POINT SAMPLING TOOL. In the POINT SAMPLING TOOL window, select SCENE_NESTBOX_LOCATIONS from the drop down menu in the LAYER CONTAINING SAMPLING POINTS section of the tool window. Then, in the LAYERS WITH FIELDS/BANDS TO GET VALUES FROM section, select the following fields/layers:

SCENE_NESTBOX_LOCATIONS: BOX_NUMBER (SOURCE POINT)
SCENE_NESTBOX_LOCATIONS: LATITUDE (SOURCE POINT)
SCENE_NESTBOX_LOCATIONS: LONGITUDE (SOURCE POINT)
SCENE_NESTBOX_LOCATIONS: BT_OCC (SOURCE POINT)
ELEVATION_MASKED: BAND 1 (RASTER)

Once you have selected all these fields and layers, enter C:/QGIS_R_WORKBOOK/ BLUE_TIT_OCCUPANCY_WITH_ELEVATION into the OUTPUT POINT VECTOR LAYER section and then click on the FIELDS tab. In the FIELDS tab, you will find a list of the fields and data layers you have selected, and a NAME field. This NAME field allows you to provide new names for the fields for each variable in the new attribute table that will be created by this tool. For SCENE_NESTBOX_LOCATIONS: BOX_NUMBER (SOURCE POINT), enter the name BOX_NUMBER (in uppercase letters). For SCENE_NESTBOX_LOCATIONS: LATITUDE (SOURCE POINT), enter the name LATITUDE. For SCENE_NESTBOX_LOCATIONS: LONGITUDE (SOURCE POINT), enter the name LONGITUDE. For SCENE_NESTBOX_ LOCATIONS: BT_OCC (SOURCE POINT), enter the name OCCUPIED. For ELEVATION_ MASKED: BAND 1 (RASTER), enter the name ELEVATION. Finally, click on the OK button to run this tool. When the tool has finished running, click on CLOSE to close the POINT SAMPLING TOOL window. **NOTE:** The new field names must be entered exactly as written here, with the all the letters in uppercase, for the R code provided later in this book to work properly If you get an error message about projection/coordinate systems when you try to run this tool, return to page 61 and check that the

Exercise Three: Linking Spatial Data Sets In QGIS And Conducting Basic Descriptive Analyses In R

appropriate projection/coordinate system as been assigned to the raster data layers you created in exercise two before you carry on with this exercise.

In the TABLE OF CONTENTS window, right click on BLUE_TIT_OCCUPANCY_ WITH_ELEVATION and select OPEN ATTRIBUTE TABLE. The contents of the TABLE window should look like this:

	BOX_NUMBER	LATITUDE	LONGITUDE	OCCUPIED	ELEVATION
0	138	56.12643700000...	-4.61782100000...	0	23.27828
1	139	56.12641253999...	-4.61816079000...	0	21.50202
2	141	56.12621899999...	-4.61796500000...	0	20.90909
3	144	56.12598299999...	-4.61698600000...	0	10.00000
4	143	56.12606799999...	-4.61688500000...	0	15.00000
5	142	56.12621599999...	-4.61699800000...	1	20.00000
6	137	56.12642499999...	-4.61700800000...	0	21.58028
7	33	56.13024099999...	-4.61416500000...	0	30.82850
8	30	56.13059400000...	-4.61469500000...	0	31.48324
9	24	56.13129000000...	-4.61525100000...	1	22.87749
10	48	56.12906499999...	-4.61481100000...	1	25.00000
11	13	56.13031399999...	-4.61620800000...	0	42.50000
12	14	56.13051600000...	-4.61645000000...	0	40.00000
13	19	56.13106500000...	-4.61664700000...	0	32.61233
14	301	56.13180899999...	-4.61724900000...	1	32.92912
15	44	56.13033200000...	-4.61723800000...	1	60.00000
16	4	56.12951600000...	-4.61551600000...	0	40.00000
17	3	56.12926999999...	-4.61511000000...	1	32.50000
18	300	56.13213900000...	-4.61671900000...	1	25.14756
19	302	56.13208499999...	-4.61702100000...	0	28.33333
20	308	56.13259999999...	-4.61738300000...	1	25.85823

As you will see, this 'big table' contains one row for each nest box. The field called BOX_NUMBER contains the unique identification number for each nest box, the LATITUDE and LONGITUDE fields contain the positional information for it, the OCCUPIED field contain information about whether a box was occupied the target species (in this case, blue tits) in a particular breeding season, while the ELEVATION field contains the information about the height of the land above sea level (in metres) for each nest box location. Thus, you have created a table which is in the format required for most statistical software packages and tests.

Exercise Three: Linking Spatial Data Sets In QGIS And Conducting Basic Descriptive Analyses In R

However, before you can start analysing the data in this table, you need to add a new field that contains information on the elevation category for each nest box. This will be used as a variable when creating a graph of occupancy later in this exercise. To do this, first in the TABLE window for the BLUE_TIT_OCCPANCY_WITH_ELEVATION layer, click on the TOGGLE EDITING MODE button (it has a yellow pencil on it). Next, click on the NEW COLUMN button (which is the second button from the right at the top of the TABLE window), and in the window then opens, enter EL_CAT for NAME (which is short for elevation category), select TEXT (STRING) for TYPE and enter 10 for WIDTH. After you have done this, click OK to create your new field. When you first add a new field, it will be empty. To fill it with the appropriate values, first open the SELECT FEATURE USING AN EXPRESSION tool by clicking on its button at the top of the TABLE window (it as a yellow square and a symbol that looks like a letter E on it). In the tool window that opens, enter the following expression:

"ELEVATION" > 0 AND "ELEVATION" <= 10

Now click the SELECT button. This will select all the data with elevation values between 0 and 10. Once these data have been selected, close the SELECT FEATURES USING AN EXPRESSION window and then click on the FIELD CALCULATOR button at the top of the TABLE window (it has a picture of an abacus on it). In the FIELD CALCULATOR window that opens, select ONLY UPDATE 9 SELECTED FEATURES at the top, and then UPDATE EXISTING FIELD. In the drop down menu directly below this, select EL_CAT, and then enter the text '0-10' in the EXPRESSION window (making sure you remember to include the single quotation marks around the text as well as the text itself) before clicking the OK button. This will add the text 0-10 to the EL_CAT field for all nest boxes with an elevation value between 0 and 10. You can now close the FIELD CALCULATOR window.

Re-open the SELECT BY EXPRESSION window and repeat this process to select the nest boxes with elevation values between 10 and 20, and fill them in with the text '10-20'. Once you have done this, do the same for nest boxes with an elevation between 20 and 30, and fill the EL_CAT field with the text '20-30', between 30 and 40 with the text '30-40', between 40 and 50 with the text '40-50' and greater than 50 with a the text '50 or more'.

Exercise Three: Linking Spatial Data Sets In QGIS And Conducting Basic Descriptive Analyses In R

For this last one, use the expression "ELEVATION" > 50 to select the required data. Once you have done this, click on the UNSELECT ALL button at the top of the TABLE window (it is beside the SELECT FEATURES USING AN EXPRESSION button), the SAVE EDITS button (it has a picture of a diskette on it) and, finally, the TOGGLE EDITING MODE button. The contents of your TABLE window should now look like this:

	BOX_NUMBER	LATITUDE	LONGITUDE	OCCUPIED	ELEVATION	EL_CAT
0	138	56.12643700000...	-4.61782100000...	0	23.27828	20-30
1	139	56.12641253999...	-4.61816079000...	0	21.50202	20-30
2	141	56.12621899999...	-4.61796500000...	0	20.90909	20-30
3	144	56.12598299999...	-4.61698600000...	0	10.00000	0-10
4	143	56.12606799999...	-4.61688500000...	0	15.00000	10-20
5	142	56.12621599999...	-4.61699800000...	1	20.00000	10-20
6	137	56.12642499999...	-4.61700800000...	0	21.58028	20-30
7	33	56.13024099999...	-4.61416500000...	0	30.82850	30-40
8	30	56.13059400000...	-4.61469500000...	0	31.48324	30-40
9	24	56.13129000000...	-4.61525100000...	1	22.87749	20-30
10	48	56.12906499999...	-4.61481100000...	1	25.00000	20-30
11	13	56.13031399999...	-4.61620800000...	0	42.50000	40-50
12	14	56.13051600000...	-4.61645000000...	0	40.00000	30-40
13	19	56.13106500000...	-4.61664700000...	0	32.61233	30-40
14	301	56.13180899999...	-4.61724900000...	1	32.92912	30-40
15	44	56.13033200000...	-4.61723800000...	1	60.00000	50 or more
16	4	56.12951600000...	-4.61551600000...	0	40.00000	30-40
17	3	56.12926999999...	-4.61511000000...	1	32.50000	30-40
18	300	56.13213900000...	-4.61671900000...	1	25.14756	20-30
19	302	56.13208499999...	-4.61702100000...	0	28.33333	20-30
20	308	56.13259999999...	-4.61738300000...	1	25.85823	20-30

You can now close the TABLE window and, if it is still open, the SELECT BY EXPRESSION window. In the next step in this exercise, you start analysing the contents of the attribute table of the BLUE_TIT_OCCUPANCY_WITH_ELEVATION data layer using the analysis tools available in R to look at the relationship between nest box occupancy and elevation in one particular breeding season.

Exercise Three: Linking Spatial Data Sets In QGIS And Conducting Basic Descriptive Analyses In R

STEP 2: CREATE A HISTOGRAM SHOWING THE DISTRIBUTION OF ALL NEST BOXES IN RELATION TO ELEVATION USING R:

In this step, you will conduct the first analysis of your data in R. However, before you can do this, you will need to create a workspace in R called NEST_BOX_DATA. To do this, open R and in the WORKSPACE window, click on the FILE menu and select SAVE AS. In the SAVE IMAGE AS window that opens, browse to the folder C:/QGIS_R_WORKBOOK and save your workspace as NEST_BOX_DATA.RDATA.

For the exercises in this workbook, you will copy the data from an attribute table of a data layer to the clipboard of your computer's operating system and then import it into R using the `read.table` command. If, for any reason, this option doesn't work for you, you can also read the contents of an attribute table directly into R. This can be done using the `read.dbf` command. This tool is found in the Foreign library. This means that in order for you to be able to use it, you will need to load this library into your R workspace. To do this, enter the following command into the R WORKSPACE window:

```
library(foreign)
```

Once you have this library installed in your workspace, you can read the attribute table of a data layer into R using the `read.dbf` command. For example, to load the contents of the attribute table for the data layer called BLUE_TIT_OCCUPANCY_WITH_ELEVATION (created in step 1 of this exercise), you would use the following command:

```
BLUE_TIT_OCCUPANCY_WITH_ELEVATION<-read.dbf(file =
"C:/QGIS_R_WORKBOOK/BLUE_TIT_OCCPANCY_WITH_ELEVATION.DBF",
                     as.is = FASLE)
```

In order to conduct your first analysis in R (which will be to create a frequency distribution showing the number of nest boxes in different elevation categories), work through the flow diagram on the next page. **NOTE:** If you have not used the exact naming protocol for your fields outlined in step 1 of this exercise (including the correct use of uppercase letters), you will need to adapt the R code provided to account for these differences otherwise it will not work properly.

Exercise Three: Linking Spatial Data Sets In QGIS And Conducting Basic Descriptive Analyses In R

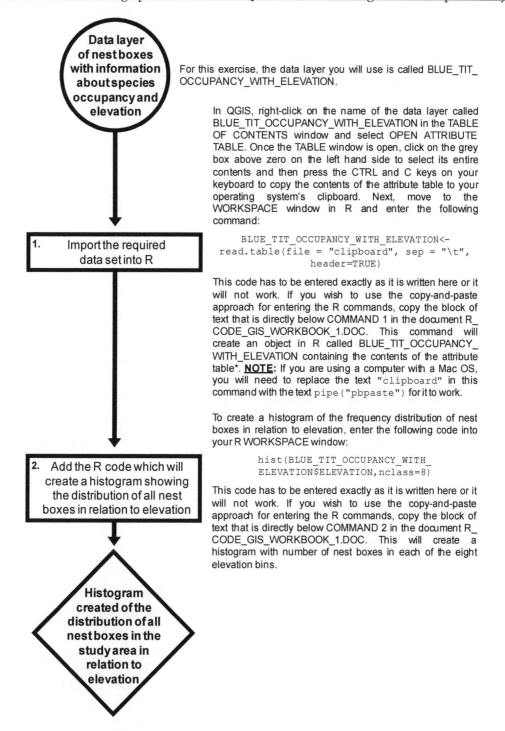

Data layer of nest boxes with information about species occupancy and elevation

For this exercise, the data layer you will use is called BLUE_TIT_OCCUPANCY_WITH_ELEVATION.

In QGIS, right-click on the name of the data layer called BLUE_TIT_OCCUPANCY_WITH_ELEVATION in the TABLE OF CONTENTS window and select OPEN ATTRIBUTE TABLE. Once the TABLE window is open, click on the grey box above zero on the left hand side to select its entire contents and then press the CTRL and C keys on your keyboard to copy the contents of the attribute table to your operating system's clipboard. Next, move to the WORKSPACE window in R and enter the following command:

```
BLUE_TIT_OCCUPANCY_WITH_ELEVATION<-
read.table(file = "clipboard", sep = "\t",
           header=TRUE)
```

1. Import the required data set into R

This code has to be entered exactly as it is written here or it will not work. If you wish to use the copy-and-paste approach for entering the R commands, copy the block of text that is directly below COMMAND 1 in the document R_CODE_GIS_WORKBOOK_1.DOC. This command will create an object in R called BLUE_TIT_OCCUPANCY_WITH_ELEVATION containing the contents of the attribute table*. **NOTE:** If you are using a computer with a Mac OS, you will need to replace the text `"clipboard"` in this command with the text `pipe("pbpaste")` for it to work.

To create a histogram of the frequency distribution of nest boxes in relation to elevation, enter the following code into your R WORKSPACE window:

```
hist(BLUE_TIT_OCCUPANCY_WITH_
ELEVATION$ELEVATION,nclass=8)
```

2. Add the R code which will create a histogram showing the distribution of all nest boxes in relation to elevation

This code has to be entered exactly as it is written here or it will not work. If you wish to use the copy-and-paste approach for entering the R commands, copy the block of text that is directly below COMMAND 2 in the document R_CODE_GIS_WORKBOOK_1.DOC. This will create a histogram with number of nest boxes in each of the eight elevation bins.

Histogram created of the distribution of all nest boxes in the study area in relation to elevation

*If, when your run the `read.table` command in R, you get an error message containing the text "*Incomplete final line found ...*", return to QGIS, copy the contents of the attribute table to your clipboard again, and then re-run Command 1 without copy-and-pasting it into your R Workspace again. One way to do this is to use the UP ARROW key on your keyboard to bring the last piece of code you ran back on to the command line in your R WORKSPACE window before pressing the ENTER key to run it again.

Once this R command has finished running, an R GRAPHICS window will open showing the frequency distribution histogram created by this code. Its contents should look like this:

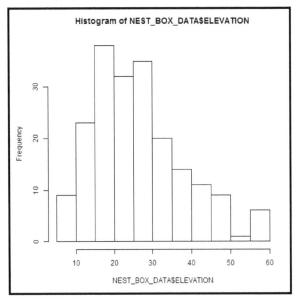

This histogram shows the distribution of all available nest boxes across eight equal elevation categories, regardless of whether or not they were occupied by blue tits, the main target species for the analyses in this exercise. If you do not get a histogram like this, check that the data have been imported correctly and that the names you have used for the fields in the attribute table of the data layer BLUE_TIT_OCCUPANCY_WITH_ELEVATION are consistent with the R code you entered in stage two of the above flow diagram (see Appendix II for details of how to do this).

NOTE: If you wished to have a different number of categories in your histogram, you would change the number after `nclass=` in the `hist` command used to create a histogram in R.

STEP 3: CREATE A HISTOGRAM SHOWING ONLY THE DISTRIBUTION OF OCCUPIED NEST BOXES IN RELATION TO ELEVATION USING R:

In this step, you will modify the R code you used in step 2 to create a new histogram that shows the distribution of only those nest boxes occupied by the targets species (in this case, blue tits) in relation to elevation. To do this, work through the following flow diagram

Exercise Three: Linking Spatial Data Sets In QGIS And Conducting Basic Descriptive Analyses In R

(**NOTE:** Since this involves working with the same data set, you do not need to import the contents of the attribute table into again):

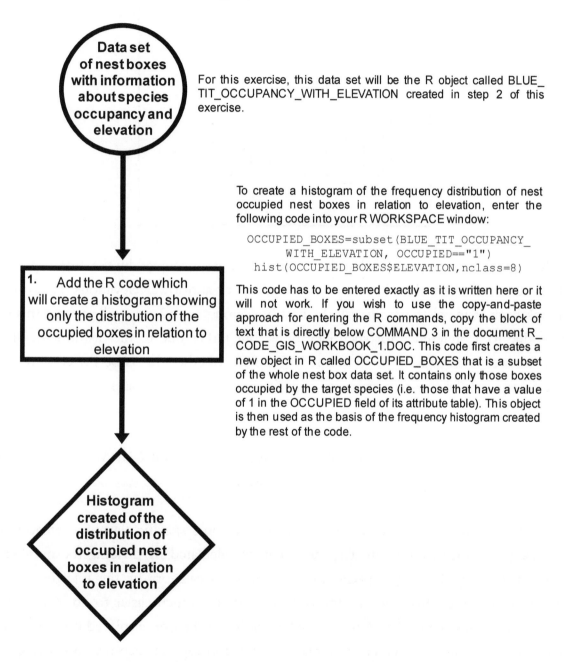

For this exercise, this data set will be the R object called BLUE_TIT_OCCUPANCY_WITH_ELEVATION created in step 2 of this exercise.

To create a histogram of the frequency distribution of nest occupied nest boxes in relation to elevation, enter the following code into your R WORKSPACE window:

```
OCCUPIED_BOXES=subset(BLUE_TIT_OCCUPANCY_
    WITH_ELEVATION, OCCUPIED=="1")
hist(OCCUPIED_BOXES$ELEVATION,nclass=8)
```

This code has to be entered exactly as it is written here or it will not work. If you wish to use the copy-and-paste approach for entering the R commands, copy the block of text that is directly below COMMAND 3 in the document R_CODE_GIS_WORKBOOK_1.DOC. This code first creates a new object in R called OCCUPIED_BOXES that is a subset of the whole nest box data set. It contains only those boxes occupied by the target species (i.e. those that have a value of 1 in the OCCUPIED field of its attribute table). This object is then used as the basis of the frequency histogram created by the rest of the code.

Once this R command has finished running, an R GRAPHICS window will open showing the frequency distribution histogram created by this code. Its contents should look like the image at the top of the next page.

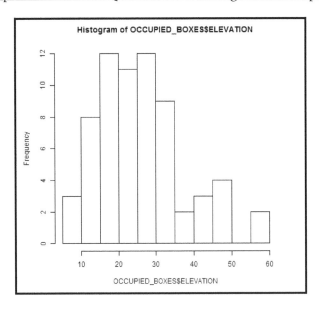

This histogram appears to show that the number of boxes occupied by the target species is highest at intermediate elevations and lowest at higher and lower elevations. However, if you look back at the histogram for all the nest boxes on page 72, you will see that this may not be due to a preference for such elevations, but rather due to the distribution of the boxes themselves in relation to elevation. To explore whether this is be the case, you can create a graph of the rate of occupancy in different elevation categories. This will be done in step 4.

STEP 4: CREATE A BAR GRAPH SHOWING THE OCCUPANCY RATE IN DIFFERENT ELEVATION CATEGORIES USING R:

An occupancy rate is the percentage of available boxes in a specific elevation category that are actually occupied by the target species. This is calculated by dividing the number of boxes occupied by the target species in each elevation category by the total number of available boxes in it. This type of summary calculation is much easier to do in R than in QGIS. Once it has been calculated, this information can then be displayed on a bar graph to allow you to examine whether there may be a relationship between nest box occupancy by the target species and elevation while taking nest box availability at different elevations into account. This can be done by working through the flow diagram provided on the next page.

Exercise Three: Linking Spatial Data Sets In QGIS And Conducting Basic Descriptive Analyses In R

Data set of nest boxes with information about species occupancy and elevation

For this exercise, this data set will be the R object called BLUE_TIT_OCCUPANCY_WITH_ELEVATION created in step 2 of this exercise.

1. Add the R code which will create a bar graph showing the occupancy rate of boxes in relation to elevation

To create a bar graph of nest box occupancy rates by blue tits in relation to elevation, enter the following code into your R WORKSPACE window:

```
NESTBOX_OCCUPANCY=aggregate(BLUE_TIT_
OCCUPANCY_WITH_ELEVATION$OCCUPIED,list(NEST_
    BOX_DATA$EL_CAT),sum)
colnames(NESTBOX_OCCUPANCY)=c("EL_CAT",
    "OCCUPIED")
NESTBOX_OCCUPANCY$N=table(BLUE_TIT_
    OCCUPANCY_WITH_ELEVATION$EL_CAT)
NESTBOX_OCCUPANCY$OCCUPANCY=((NESTBOX_
OCCUPANCY$OCCUPIED/NESTBOX_OCCUPANCY$N)*100)
    barplot(NESTBOX_OCCUPANCY$OCCUPANCY,
xlab="ELEVATION CATEGORIES", ylab="OCCUPANCY
RATES", names.arg=NESTBOX_OCCUPANCY$EL_CAT)
```

This code has to be entered exactly as it is written here or it will not work. If you wish to use the copy-and-paste approach for entering the R commands, copy the block of text that is directly below COMMAND 4 in the document R_CODE_GIS_WORKBOOK_1.DOC.

This code creates a table in R (called NEST_BOX_OCCUPANCY) which is then used to calculate the occupancy rates for each elevation category in the EL_CAT field. The information from this table is then used as the basis of the bar graph created by the `barplot` command. In addition, within the `barplot` code, names are specified for the X axis (using the code `xlab`), the Y axis (using the code `ylab`) and for the categories to be displayed on the X axis (using the `names.arg` code).

Bar graph created of the occupancy rate of nest boxes in the study area in relation to elevation

Exercise Three: Linking Spatial Data Sets In QGIS And Conducting Basic Descriptive Analyses In R

Once this R command has finished running, an R GRAPHICS window will open showing the bar graph created by this code. Its contents should look like this:

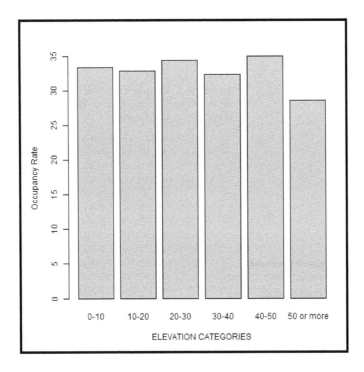

This bar graph shows the percentage of nest boxes occupied by blue tits in each elevation category. It suggests that there is no clear relationship between nest box occupancy by blue tits and elevation when the availability of nest boxes at different elevations is taken into account.

STEP 5: CALCULATE THE AVERAGE ELEVATION FOR OCCUPIED AND UNOCCUPIED NEST BOXES USING R:

As well as creating graphs from your data using R, you can also extract summary statistics such averages, standard deviations, and minimum and maximum values. To show you how to do this, in this step you will calculate the average elevation value for nest boxes that were occupied by blue tits and those that were not. To do this, work through the flow diagram that starts at the top of the next page.

Exercise Three: Linking Spatial Data Sets In QGIS And Conducting Basic Descriptive Analyses In R

Data set of nest boxes with information about species occupancy and elevation

For this exercise, this data set will be the R object called BLUE_TIT_OCCUPANCY_WITH_ELEVATION created in step 2 of this exercise.

To calculate the average elevation of the nest boxes occupied by blue tits, and those that were not, enter the following code into your R WORKSPACE window:

```
OCCUPIED_BOXES=subset(BLUE_TIT_
OCCUPANCY_WITH_ELEVATION, OCCUPIED=="1")
   mean(OCCUPIED_BOXES$ELEVATION)
UNOCCUPIED_BOXES=subset(BLUE_TIT_
OCCUPANCY_WITH_ELEVATION, OCCUPIED=="0")
   mean(UNOCCUPIED_BOXES$ELEVATION)
```

1. Add the R code to calculate the mean elevation of occupied and unoccupied boxes in the data set

This code has to be entered exactly as it is written here or it will not work. If you wish to use the copy-and-paste approach for entering the R commands, copy the block of text that is directly below COMMAND 5 in the document R_CODE_WORKBOOK_1.DOC.

The code first creates a new object in R called OCCUPIED_BOXES which contains just those nest boxes that were occupied by nest boxes (which is designated by a value of 1 in the field called OCCUPIED in the BLUE_TIT_OCCUPANCY_WITH_ELEVATION data set). This new R object is the used to calculate the average of the values in the field called ELEVATION using the mean command. This process in then repeated to create a second new R object containing just the unoccupied nest boxes (called UNOCCUPIED_BOXES) and then calculate an average of the elevation values for these boxes.

Means of elevation calculated for occupied and unoccupied nest boxes

Exercise Three: Linking Spatial Data Sets In QGIS And Conducting Basic Descriptive Analyses In R

Once this R command has finished running, the contents of your R WORKSPACE window should look like this:

```
> OCCUPIED_BOXES=subset(BLUE_TIT_OCCUPANCY_WITH_ELEVATION, OCCUPIED=="1")
> mean(OCCUPIED_BOXES$ELEVATION)
[1] 26.61863
> UNOCCUPIED_BOXES=subset(BLUE_TIT_OCCUPANCY_WITH_ELEVATION, OCCUPIED=="0")
> mean(UNOCCUPIED_BOXES$ELEVATION)
[1] 26.91886
>
```

The average elevation for nest boxes occupied by blue tits can be found directly below the `mean(OCCUPIED_BOXES$ELEVATION)` command, while the average elevation for nest boxes not occupied by blue tits can be found directly below the `mean(UNOCCUPIED_BOXES$ELEVATION)` command. As you can see, these two values look very similar, but you cannot necessarily conclude that there is no difference between them without doing a statistical analysis. This will be done using a t-test in step 6.

STEP 6: TEST WHETHER THERE IS A SIGNIFICANT DIFFERENCE IN THE MEAN ELEVATION VALUES FOR OCCUPIED AND UNOCCUPIED NEST BOXES USING R:

To test whether there is a significant difference between the mean elevations of occupied and unoccupied nest boxes, you will run a t-test on your data set. When doing this, it will be assumed that you have already tested your data to check whether they are consistent with the requirements of a t-test (such having equal variance in the elevation values for occupied and unoccupied nest boxes). However, if you were doing this on your own data, you would need to conduct these checks on your data before running this (or any other) statistical test. To run a t-test on your nest box data set, work through the flow diagram provided on the next page.

Exercise Three: Linking Spatial Data Sets In QGIS And Conducting Basic Descriptive Analyses In R

Data set of nest boxes with information about species occupancy and elevation

For this exercise, this data set will be the R object called BLUE_TIT_OCCUPANCY_WITH_ELEVATION created in step 2 of this exercise.

To test whether there is a significant difference in the average elevation of occupied and unoccupied nest boxes, enter the following code into your R WORKSPACE window:

```
t.test(ELEVATION ~ OCCUPIED, data = BLUE_
TIT_OCCUPANCY_WITH_ELEVATION, var.equal =
                       TRUE)
```

1. Add the R code to run a t-test on the average elevation of occupied and unoccupied boxes

This code has to be entered exactly as it is written here or it will not work. If you wish to use the copy-and-paste approach for entering the R commands, copy the block of text that is directly below COMMAND 6 in the document R_CODE_GIS_WORKBOOK_1.DOC.

This code uses the `t.test` command to compare values contained in one field of an R object (this case, the ELEVATION field of the R object called BLUE_TIT_OCCUPANCY_WITH_ELEVATION) based on the groupings contained in another one (in this case, the OCCUPIED field which contains a value of 1 for nest boxes occupied by blue tits, and a value of 0 for those that were not).

T-test conducted to test for a significant difference in the mean elevation between occupied and unoccupied nest boxes

Exercise Three: Linking Spatial Data Sets In QGIS And Conducting Basic Descriptive Analyses In R

Once this R command has finished running, the contents of your R WORKSPACE window should look like this:

```
> t.test(ELEVATION ~ OCCUPIED, data = BLUE_TIT_OCCUPANCY_WITH_ELEVATIO

        Two Sample t-test

data:  ELEVATION by OCCUPIED
t = 0.16937, df = 196, p-value = 0.8657
alternative hypothesis: true difference in means is not equal to 0
95 percent confidence interval:
 -3.195685  3.796133
sample estimates:
mean in group 0 mean in group 1
       26.91886        26.61863

>
```

The results of this t-test show that the value of t for this comparison is 0.16937, while the degrees of freedom (d.f.) are 196. This gives a p-value of 0.8657, which exceeds the 0.05 threshold that is usually used to determine whether, or not, two samples have significantly different means. Thus, you can conclude from this that there is no significant difference in the mean value of elevation between those boxes that were occupied by blue tits in the particular breeding season under investigation and those that were not. This is what would have been expected given the occupancy rate graph generated in step 4 and the values for the mean elevations generated in step 5.

STEP 7: CALCULATE OTHER SUMMARY STATISTICS FOR OCCUPIED AND UNOCCUPIED NEST BOXES USING R:

There are many other summary statistics that you can calculated and compare for your data sets. In this final step in this exercise, you will generate tables with a range of summary statistics for those nest boxes occupied by blue tits and those which were not. These will include the sample size, the number of unique values, the minimum value, the maximum value, the range of values, the mean, the median and the standard deviation. To do this, work through the flow diagram that starts on the next page.

Exercise Three: Linking Spatial Data Sets In QGIS And Conducting Basic Descriptive Analyses In R

Data set of nest boxes with information about species occupancy and elevation

For this exercise, this data set will be the R object called BLUE_TIT_OCCUPANCY_WITH_ELEVATION created in step 2 of this exercise.

1. Add the R code to create a table of summary statistics of elevation values for occupied nest boxes

To create a table of summary statistics for the nest boxes occupied by blue tits, enter the following code into your R WORKSPACE window:

```
OCCUPIED_BOXES=subset(BLUE_TIT_
OCCUPANCY_WITH_ELEVATION, OCCUPIED=="1")
      SUMMARY_STATISTICS_1<-
data.frame(rbind(length(OCCUPIED_BOXES$
             ELEVATION),
length(unique(OCCUPIED_BOXES$ELEVATION)),
    min(OCCUPIED_BOXES$ELEVATION),
    max(OCCUPIED_BOXES$ELEVATION),
    max(OCCUPIED_BOXES$ELEVATION)-
    min(OCCUPIED_BOXES$ELEVATION),
   mean(OCCUPIED_BOXES$ELEVATION),
  median(OCCUPIED_BOXES$ELEVATION),
    sd(OCCUPIED_BOXES$ELEVATION)),
    row.names=c("Count:","Unique
  values:","Minimum value:","Maximum
 value:","Range:","Mean value:","Median
   value:","Standard deviation:"))
colnames(SUMMARY_STATISTICS_1)<-c("ELEVATION
       OF OCCUPIED BOXES")
       SUMMARY_STATISTICS_1
```

This code has to be entered exactly as it is written here or it will not work. If you wish to use the copy-and-paste approach for entering the R commands, copy the block of text that is directly below COMMAND 7 in the document R_CODE_GIS_WORKBOOK_1.DOC.

This code creates a new object from the BLUE_TIT_OCCUPANCY_WITH_ELEVATION data set called OCCUPIED_BOXES which contains just the data from the nest boxes occupied by blue tits (as indicated by having a value of 1 in the field called OCCUPIED). A second new R object called SUMMARY_STATISTICS_1 is then created. This is a table which is then filled with summary statistics for the nest boxes occupied by blue tits using various calculation functions from R.

Exercise Three: Linking Spatial Data Sets In QGIS And Conducting Basic Descriptive Analyses In R

2. Add the R code to create a table of summary statistics of elevation values for unoccupied nest boxes

To calculate a table of summary statistics for the nest boxes not occupied by blue tits, enter the following code into your R WORKSPACE window:

```
UNOCCUPIED_BOXES=subset(BLUE_TIT_
  OCCUPANCY_WITH_ELEVATION, OCCUPIED=="0")
      SUMMARY_STATISTICS_2<-
   data.frame(rbind(length(UNOCCUPIED_BOXES$
             ELEVATION),
length(unique(UNOCCUPIED_BOXES$ELEVATION)),
       min(UNOCCUPIED_BOXES$ELEVATION),
       max(UNOCCUPIED_BOXES$ELEVATION),
       max(UNOCCUPIED_BOXES$ELEVATION)-
       min(UNOCCUPIED_BOXES$ELEVATION),
       mean(UNOCCUPIED_BOXES$ELEVATION),
      median(UNOCCUPIED_BOXES$ELEVATION),
       sd(UNOCCUPIED_BOXES$ELEVATION)),
        row.names=c("Count:","Unique
    values:","Minimum value:","Maximum
  value:","Range:","Mean value:","Median
     value:","Standard deviation:"))
colnames(SUMMARY_STATISTICS_2)<-c("ELEVATION
        OF UNOCCUPIED BOXES")
        SUMMARY_STATISTICS_2
```

This code has to be entered exactly as it is written here or it will not work. If you wish to use the copy-and-paste approach for entering the R commands, copy the block of text that is directly below COMMAND 8 in the document R_CODE_GIS_WORKBOOK_1.DOC.

This code creates a new object from the BLUE_TIT_OCCUPANCY_WITH_ELEVATION data set called UNOCCUPIED_BOXES which contains just the data from the nest boxes not occupied by blue tits (as indicated by having a value of 0 in the field called OCCUPIED). A second new R object called SUMMARY_STATISTICS_2 is then created. This is a table which is then filled with summary statistics for the nest boxes not occupied by blue tits using various calculation functions from R.

Tables of summary statistics created for occupied and unoccupied nest boxes

Once you have created the tables of summary statistics using this code, you can then view them using the `View` command (**NOTE:** Unlike most other R commands, the `View` command starts with an uppercase letter). To view the table of summary statistics for the nest boxes occupied by blue tits, enter the following command into your R WORKSPACE:

```
View(SUMMARY_STATISTICS_1)
```

The VIEWER window that opens, should look like this:

	row.names	ELEVATION OF OCCUPIED BOXES
1	Count:	66.00000
2	Unique values:	48.00000
3	Minimum value:	8.43260
4	Maximum value:	60.00000
5	Range:	51.56740
6	Mean value:	26.61863
7	Median value:	25.00000
8	Standard deviation:	11.73875

To view the table of summary statistics for the nest boxes not occupied by blue tits, enter the following command into your R WOKSPACE:

```
View(SUMMARY_STATISTICS_2)
```

The VIEWER window that opens, should look like this:

	row.names	ELEVATION OF UNOCCUPIED BOXES
1	Count:	132.00000
2	Unique values:	87.00000
3	Minimum value:	8.50000
4	Maximum value:	57.73448
5	Range:	49.23448
6	Mean value:	26.91886
7	Median value:	25.00000
8	Standard deviation:	11.76814

Exercise Three: Linking Spatial Data Sets In QGIS And Conducting Basic Descriptive Analyses In R

Optional extra:

If you wish to have more practice at linking your data to environmental variables in QGIS and running basic descriptive analyses using R, you can repeat this exercise using a second species, the great tit, as the target species and investigate the relationship between great tit nest box occupancy and elevation. In order to do this, you will need to have completed the optional extra for exercise one.

Alternatively, you can repeat this exercise using the environmental variable slope instead of the elevation. When you do this, make sure that you use different names for any data layers you generate to ensure you do not over-write any of your existing data sets, and remember that you will need to update the field and file names in the R code before you can run the analyses on your new data sets.

--- Chapter Six ---

Exercise Four: Conducting Linear Regressions With Generalised Linear Modelling (GLM) Using QGIS And R

In exercise three, you carried out some basic descriptive analyses of the relationship between nest box occupancy by blue tits and elevation in the data set being used for the exercises in this workbook. In this exercise, you will carry out more detailed analyses. These will be a linear regression of the relationship between nest box occupancy and elevation, and a second, more complex, linear regression which will compare nest box occupancy not just to elevation, but also to slope, hillshade, the distance to the edge of Loch Lomond and the distance to the edge of the woodland patch where the nest boxes are sited. Both of these regression analysis will be done using the GLM function in R, which allows you to conduct linear regressions using generalised linear modelling.

The starting point for this exercise is the GIS project created in exercise three. Open QGIS. Once it is open, click on the PROJECT menu and select OPEN. When the OPEN window appears, browse to the location where your project is saved (C:\QGIS_R_WORBOOK), select it and click OPEN. After you have opened the GIS project called EXERCISE_THREE, the first thing you need to do is save it under a new name. This is because you do not want to alter the contents of the original project, you just want to base your new one on it since this saves you having to add all the data layers again, and also having to reset the projection/coordinate system of the data frame. To save the project under a new name, click on PROJECT on the main menu bar, and select SAVE AS. For this exercise, save it as EXERCISE_FOUR in the C:\QGIS_R_WORKBOOK folder.

Exercise Four: Conducting Linear Regressions With GLMs Using QGIS And R

The data layers that you will need for this exercise are the feature data layer BLUE_TIT_OCCUPANCY_WITH_ELEVATION, and the raster data layers SLOPE, HILLSHADE, DIST_LOCH and DIST_EDGE. Move these data layers to the top of your TABLE OF CONTENTS window until the top part of it looks like this (other layers will still be included below these ones):

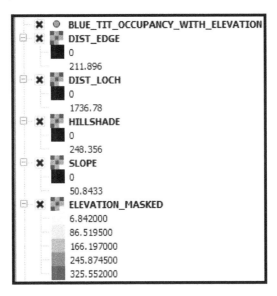

Now make sure that the four additional environmental raster data layers (SLOPE, HILLSHADE, DIST_LOCH and DIST_EDGE) are set to display in the MAP window. Its contents should now look like this:

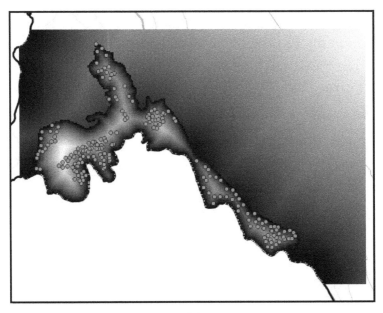

STEP 1: CONDUCT A LINEAR REGRESSION OF THE RELATIONSHIP BETWEEN NEST BOX OCCUPANCY AND ELEVATION USING R:

In exercise three, you conducted a basic t-test in R to compare the mean elevation between occupied and unoccupied nest boxes. In this step, you will conduct a more detailed analysis. This will be a logistic, or binomial, linear regression of the relationship between nest box occupancy and elevation. A logistic regression is a regression where the response variable (in this case, whether a particular nest box was occupied or not) has a binomial distribution. In R, the easiest way to conduct a logistic regression is using the GLM (generalised linear modelling) function. This analysis can be done using the contents of attribute table of the feature data layer called BLUE_TIT_OCCUPANCY_WITH_ELEVATION which was created at the start of exercise three. This is because it has one field that contains information about whether each nest box was occupied or unoccupied by the target species (called OCCUPIED), and another than contains the associated elevation data (called ELEVATION). These data have already been imported into R in step 2 of exercise three, where they were used to create an R object called BLUE_TIT_OCCUPANCY_WITH_ELEVATION in the workspace called NEST_BOX_DATA.RDATA. It is this existing R object that will be used as the basis for the analysis you will carry out in this step. If this workspace is not already open, open the R CONSOLE window and load it into R by clicking on the FILE menu and selecting LOAD WORKSPACE.

There are two main steps in the linear regression process. The first is to create an X-Y scatter plot which shows the relationship between the response variable (in this case, nest box occupancy) and the explanatory variable (in this case, elevation). This scatter plot will include a line of best fit, which will give an indication of the relationship between the two variables. In the second step, a generalised linear model (GLM) is used to assess the significance of this relationship. To conduct a logistic linear regression of the relationship between nest box occupancy and elevation, work through the flow diagram that starts at the top of the next page.

Exercise Four: Conducting Linear Regressions With GLMs Using QGIS And R

Data set of nest boxes with information about species occupancy and elevation

For this exercise, this data set will be the R object called BLUE_TIT_OCCUPANCY_WITH_ELEVATION created in step 2 of exercise three.

1. Create a scatter plot of nest box occupancy against elevation

To create a scatter plot of the relationship between nest box occupancy by blue tits and elevation, enter the following code into your R WORKSPACE window:

```
plot(BLUE_TIT_OCCUPANCY_
  WITH_ELEVATION$ELEVATION, BLUE_TIT_
  OCCUPANCY_WITH_ELEVATION$OCCUPIED,
  main="OCCUPANCY VS ELEVATION",
  xlab="ELEVATION", ylab="OCCUPANCY")
abline(lm(OCCUPIED~ELEVATION, data=BLUE_
  TIT_OCCUPANCY_WITH_ELEVATION))
```

This code has to be entered exactly as it is written here or it will not work. If you wish to use the copy-and-paste approach for entering the R commands, copy the block of text that is directly below COMMAND 9 in the document R_CODE_GIS_WORKBOOK_1.DOC.

This code uses the `plot` command to create a scatter plot based on the information in the fields ELEVATION and OCCUPIED in the dataset BLUE_TIT_OCCUPANCY_WITH_ELEVATION. In addition, within the `plot` code, names are specified for the X axis (using the code `xlab`) and the Y axis (using the code `ylab`). Once this scatter plot has been created, a line of best fit between these two variables in the BLUE_TIT_OCCUPANCY_WITH_ELEVATION data set is added to it using the `abline` command.

Exercise Four: Conducting Linear Regressions With GLMs Using QGIS And R

2. Test the significance of the relationship between nest box occupancy and elevation using a linear regression

To run a linear regression on the relationship between nest box occupancy by blue tits and elevation, enter the following code into your R WORKSPACE window:

```
LOGISTIC_REGRESSION<-glm(OCCUPIED ~
ELEVATION, family=binomial(link='logit'),
data=BLUE_TIT_OCCUPANCY_WITH_ELEVATION)
     summary(LOGISTIC_REGRESSION)
```

This code has to be entered exactly as it is written here or it will not work. If you wish to use the copy-and-paste approach for entering the R commands, copy the block of text that is directly below COMMAND 10 in the document R_CODE_GIS_WORKBOOK_1.DOC.

This code uses the `glm` command to create the linear regression based on the information in the fields ELEVATION and OCCUPIED in the dataset BLUE_TIT_OCCUPANCY_WITH_ELEVATION. Within the `glm` code, the `family=binomial(link='logit')` term is used to denote that the regression being run should be a logistic regression based on binomial data (occupied vs unoccupied) rather than one based on any other possible distribution that can be used with the `glm` function. Finally, the `summary` command is used to call up and display the results of this analysis.

Linear regression conducted to test for the significance of the relationship between nest box occupancy and elevation

Exercise Four: Conducting Linear Regressions With GLMs Using QGIS And R

At the end of this step, an R GRAPHICS window will open containing an X-Y scatter plot. Its contents should look like this:

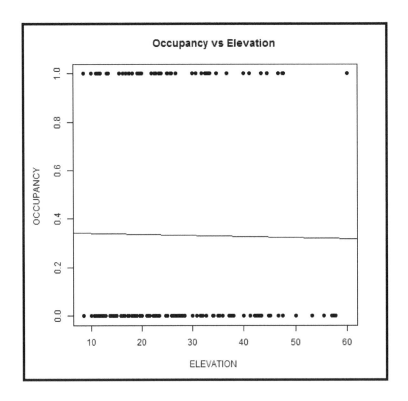

This scatter plot shows the relationship between nest box occupancy and elevation. As you can see from the line of best fit, there does not appear to be a strong relationship between these two variables. If you move back to the R WORKSPACE window, it should now be displaying the results of the GLM (which have been called up by the summary command). These results should look like the image at the top of the next page.

```
Call:
glm(formula = OCCUPIED ~ ELEVATION, family = binomial(link = "logit"),
    data = BLUE_TIT_OCCUPANCY_WITH_ELEVATION)

Deviance Residuals:
    Min       1Q   Median       3Q      Max
-0.9155  -0.9049  -0.8946   1.4736   1.5152

Coefficients:
             Estimate Std. Error z value Pr(>|z|)
(Intercept) -0.63425    0.37686  -1.683   0.0924 .
ELEVATION   -0.00220    0.01293  -0.170   0.8649
---
Signif. codes:  0 '***' 0.001 '**' 0.01 '*' 0.05 '.' 0.1 ' ' 1

(Dispersion parameter for binomial family taken to be 1)

    Null deviance: 252.06  on 197  degrees of freedom
Residual deviance: 252.03  on 196  degrees of freedom
AIC: 256.03

Number of Fisher Scoring iterations: 4

> |
```

These results show the logistic linear regression found that the relationship between nest box occupancy and elevation was not significant (p=0.8649 – highlighted here with a black box, which will not appear on your output). However, this does not necessarily mean that there is no relationship between nest box occupancy and elevation, but rather that there is no evidence of a simple relationship between the two variables. It is still possible that there is a relationship, but that it is obscured by the effects of other environmental variables, such as slope, hillshade, the distance to the edge of Loch Lomond and/or the distance to the edge of the patch of oak woodland in which the nest boxes are sited. To investigate this possibility, you can expand the regression analysis to include multiple explanatory variables. However, before you can do this, you will need to link your nest box occupancy data to these additional environmental variables. This will be done in the next step.

STEP 2: LINK YOUR NEST BOX OCCUPANCY DATA TO ADDITIONAL ENVIRONMENTAL VARIABLES IN QGIS:

In this step, you will return to QGIS to create a second 'big table' using the POINT SAMPLING TOOL plugin that will include information about slope, hillshade, the distance from Loch Lomond and the distance to the edge of the patch of oak woodland, as well as the existing information on nest box occupancy and elevation. If you have created the raster data layers of environmental variables properly, so that they all have the same extent, compatible cell sizes and are all in the same projection/coordinate system, this is surprisingly quick and easy to do. However, this is where any incompatibilities between your data layers will become apparent and you may find you have to go back and repeat earlier steps in order to get it to work properly.

To link the nest box data to the additional environmental raster data layers, first make sure that the point data layer called BLUE_TIT_OCCUPANCY_WITH_ELEVATION and the raster data layers SLOPE, HILLSHADE, DIST_LOCH and DIST_EDGE are set to display (i.e. they have a cross in the box beside their names in the TABLE OF CONTENTS window). Next, click on PLUGINS on the main menu bar and select ANALYSES> POINT SAMPLING TOOL. In the POINT SAMPLING TOOL window, select BLUE_TIT_OCCUPANCY_WITH_ELEVATION from the drop down menu in the LAYER CONTAINING SAMPLING POINTS section of the tool window. Then, in the LAYERS WITH FIELDS/BANDS TO GET VALUES FROM section, select the following fields/layers:

   BLUE_TIT_OCCUPANCY_WITH_ELEVATION: BOX_NUMBER (SOURCE POINT)
   BLUE_TIT_OCCUPANCY_WITH_ELEVATION: LATITUDE (SOURCE POINT)
   BLUE_TIT_OCCUPANCY_WITH_ELEVATION: LONGITUDE (SOURCE POINT)
   BLUE_TIT_OCCUPANCY_WITH_ELEVATION: OCCUPIED (SOURCE POINT)
   BLUE_TIT_OCCUPANCY_WITH_ELEVATION: ELEVATION (SOURCE POINT)
   BLUE_TIT_OCCUPANCY_WITH_ELEVATION: EL_CAT (SOURCE POINT)
   SLOPE: BAND 1 (RASTER)
   HILLSHADE: BAND 1 (RASTER)
   DIST_LOCH: BAND 1 (RASTER)
   DIST_EDGE: BAND 1 (RASTER)

Once you have selected all these fields and layers, enter C:/QGIS_R_WORKBOOK/ BLUE_TIT_OCCUPANCY_WITH_ALL_EGVS into the OUTPUT POINT VECTOR LAYER section and then click on the FIELDS tab. In the FIELDS tab, you will find a list of the fields and data layers you have selected, and a NAME field. This NAME field allows you to provide new names for the fields for each variable in the new attribute table that will be created by this tool. Check that the following names are being used for the relevant fields: BOX_NUMBER, LATITUDE, LONGITUDE, OCCUPIED, ELEVATION, EL_CAT, SLOPE, HILLSHADE, DIST_LOCH, DIST_EDGE (**NOTE:** The new field names must be entered exactly as written here, with the all the letters in uppercase, for the R code provided later in this book to work properly). If any of these need to be changed, change them to the appropriate name and then click on the OK button to run this tool. When the tool has finished running, click on CLOSE to close the POINT SAMPLING TOOL window.

In the TABLE OF CONTENTS window, find the data layer called BLUE_TIT_ OCCUPANCY_WITH_ALL_EGVS, click on it and drag it to the top of the list of data layers in this window. Now right click on the name of this data layer and select OPEN ATTRIBUTE TABLE. The contents of the TABLE window should look like this:

	BOX_NUMBER	LATITUDE	LONGITUDE	OCCUPIED	ELEVATION	EL_CAT	SLOPE	HILLSHADE	DIST_LOCH	DIST_EDGE
0	138	56.12643700000...	-4.61782100000...	0	23.27828	20-30	4.97327	143.91649	80.62257	80.62257
1	139	56.12641253999...	-4.61816079000...	0	21.50202	20-30	5.70668	142.98721	92.19544	92.19544
2	141	56.12621899999...	-4.61796500000...	0	20.90909	20-30	4.87828	145.77516	92.19544	92.19544
3	144	56.12598299999...	-4.61698600000...	0	10.00000	0-10	19.06871	176.63585	31.62278	31.62278
4	143	56.12606799999...	-4.61688500000...	0	15.00000	10-20	24.06850	195.37674	41.23106	41.23106
5	142	56.12621599999...	-4.61699800000...	1	20.00000	10-20	7.80331	155.47945	44.72136	50.00000
6	137	56.12642499999...	-4.61700800000...	0	21.58028	20-30	2.67440	137.21011	36.05551	36.05551
7	33	56.13024099999...	-4.61416500000...	0	30.82850	30-40	13.76794	90.15897	233.45235	36.05551
8	30	56.13059400000...	-4.61469500000...	0	31.48324	30-40	2.16119	121.04276	278.92651	50.99020
9	24	56.13129000000...	-4.61525100000...	1	22.87749	20-30	4.61178	112.82821	364.00549	70.00000
10	48	56.12906499999...	-4.61481100000...	1	25.00000	20-30	22.64382	113.54879	134.53624	128.06248
11	13	56.13031399999...	-4.61620800000...	0	42.50000	40-50	19.32568	83.41280	294.10883	58.30952
12	14	56.13051600000...	-4.61645000000...	0	40.00000	30-40	20.40451	66.36616	324.49960	60.00000
13	19	56.13106500000...	-4.61664700000...	0	32.61233	30-40	13.14117	112.02937	380.78867	50.00000
14	301	56.13180899999...	-4.61724900000...	1	32.92912	30-40	14.13227	117.36760	470.10638	80.62257
15	44	56.13033200000...	-4.61723800000...	1	60.00000	50 or more	26.76279	35.79152	336.15472	14.14214
16	4	56.12951600000...	-4.61551600000...	0	40.00000	30-40	17.87830	126.84810	198.49434	102.95630
17	3	56.12926999999...	-4.61511000000...	1	32.50000	30-40	15.22593	148.58337	164.01219	120.83046
18	300	56.13213900000...	-4.61671900000...	0	25.14756	20-30	9.02904	114.15614	488.46698	92.19544
19	302	56.13208499999...	-4.61702100000...	0	28.33333	20-30	8.64850	104.39606	487.54486	107.70329
20	308	56.13259999999...	-4.61738300000...	1	25.85823	20-30	8.83440	105.29299	550.36353	70.71068

As you will see, this new 'big table' now contains all the original fields as well as new fields for the additional environmental data layers (in this case, SLOPE, HILLSHADE, DIST_LOCH and DIST_EDGE). Thus, you have created a table which you can use to test whether any of these additional variables influence nest box occupancy. This will be done in step 4, but before you do this, you need to explore the individual relationships between nest box occupancy and the additional environmental variables. This will be done in step 3.

Exercise Four: Conducting Linear Regressions With GLMs Using QGIS And R

STEP 3: CREATE SCATTER PLOTS TO EXPLORE THE RELATIONSHIPS BETWEEN NEST BOX OCCUPANCY AND THE ADDITIONAL ENVIRONMENTAL VARIABLES USING R:

It is always worth exploring the relationships between each individual explanatory variable and the response variable before you run any statistical analysis. This can be done by creating scatter plots of these relationships with a line of best fit line added to them. You can do this in R by adapting the code used at the beginning of step 1 of this exercise to allow you create scatter plots for the relationship between nest box occupancy by blue tits and the additional environmental variables. However, before you can do this, you need to import the new data set incorporating the additional environmental variables created in step 2 of this exercise into R. This can be done by selecting the contents of the attribute table of the data layer BLUE_TIT_OCCUPANCY_WITH_ALL_EGVS and copying it to the clipboard of your operating system. It can then be imported into R by running the following command in your R WORKSPACE window:

```
BLUE_TIT_OCCUPANCY_WITH_ALL_EGVS<-read.table(file =
    "clipboard", sep = "\t", header=TRUE)
```

This code has to be entered exactly as it is written here or it will not work. If you wish to use the copy-and-paste approach for entering the R commands, copy the block of text that is directly below COMMAND 11 in the document R_CODE_GIS_WORKBOOK_1.DOC. **NOTE:** If you are using a computer with a Mac OS, you will need to replace the text `"clipboard"` in this command with the text `pipe("pbpaste")` in order for it to work. This command will create a new object in R called BLUE_TIT_OCCUPANCY_WITH_ALL_EGVS that contains not just the information on occupancy and elevation but also the information about the additional environmental data layers. If, when your run this `read.table` command, you get an error message containing the text *"Incomplete final line found ..."*, return to QGIS, copy the contents of the attribute table to your clipboard again, and then re-run this command without copy-and-pasting it into your R Workspace again. One way to do this is to use the UP ARROW key on your keyboard to bring the last piece of code you ran back on to the command line in your R WORKSPACE window before pressing the ENTER key to run it again.

Once you have created the BLUE_TIT_OCCUPANCY_WITH_ALL_EGVS object in R containing the new environmental variables, you can create a scatter plot of the relationship between occupancy and slope by entering the following command into your R WORKSPACE window:

```
plot(BLUE_TIT_OCCUPANCY_WITH_ALL_EGVS$SLOPE,
BLUE_TIT_OCCUPANCY_WITH_ALL_EGVS$OCCUPIED, main="OCCUPANCY
     VS SLOPE", xlab="SLOPE", ylab="OCCUPANCY")
    abline(lm(OCCUPIED~SLOPE, data= BLUE_TIT_OCCUPANCY_
                    WITH_ALL_EGVS))
```

If you wish to copy and paste this command into R, it is COMMAND 12 in R_CODE_GIS_WORKBOOK_1.DOC. Once this command has finished running, an R GRAPHICS window should open containing a scatter plot that looks like this:

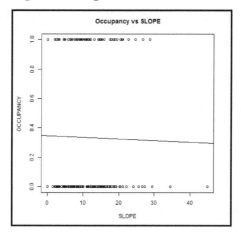

To create a scatter plot of the relationship between occupancy and hillshade using the following enter the following command into your R WORKSPACE window:

```
plot(BLUE_TIT_OCCUPANCY_WITH_ALL_EGVS$HILLSHADE,
BLUE_TIT_OCCUPANCY_WITH_ALL_EGVS$OCCUPIED, main="OCCUPANCY
   VS HILLSHADE", xlab="HILLSHADE", ylab="OCCUPANCY")
      abline(lm(OCCUPIED~HILLSHADE, data= BLUE_TIT_
                 OCCUPANCY_WITH_ALL_EGVS))
```

If you wish to copy and paste this command into R, it is COMMAND 13 in R_CODE_GIS_WORKBOOK_1.DOC. Once this command has finished running, an R GRAPHICS window should open containing a scatter plot that looks like the image at the top of the next page.

Exercise Four: Conducting Linear Regressions With GLMs Using QGIS And R

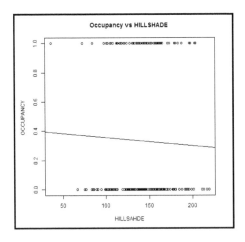

To create a scatter plot of the relationship between occupancy and distance to the edge of Loch Lomond, enter the following command into your R WORKSPACE window:

```
plot(BLUE_TIT_OCCUPANCY_WITH_ALL_EGVS$DIST_LOCH,
BLUE_TIT_OCCUPANCY_WITH_ALL_EGVS$OCCUPIED, main="OCCUPANCY
   VS DIST_LOCH", xlab="DIST_LOCH", ylab="OCCUPANCY")
      abline(lm(OCCUPIED~DIST_LOCH, data= BLUE_TIT_
                 OCCUPANCY_WITH_ALL_EGVS))
```

If you wish to copy and paste this command into R, it is COMMAND 14 in R_CODE_GIS_WORKBOOK_1.DOC. Once this command has finished running, an R GRAPHICS window should open containing a scatter plot that looks like this:

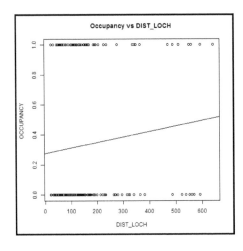

To create a scatter plot of the relationship between occupancy and the distance to the edge of the woodland where the nest boxes are sited, enter the following command into your R WORKSPACE window:

```
plot(BLUE_TIT_OCCUPANCY_WITH_ALL_EGVS$DIST_EDGE,
BLUE_TIT_OCCUPANCY_WITH_ALL_EGVS$OCCUPIED, main="OCCUPANCY
     VS DIST_EDGE", xlab="DIST_EDGE", ylab="OCCUPANCY")
     abline(lm(OCCUPIED~DIST_EDGE, data= BLUE_TIT_
              OCCUPANCY_WITH_ALL_EGVS))
```

If you wish to copy and paste this command into R, it is COMMAND 15 in R_CODE_GIS_WORKBOOK_1.DOC. Once this command has finished running, an R GRAPHICS window should open containing a scatter plot that looks like this:

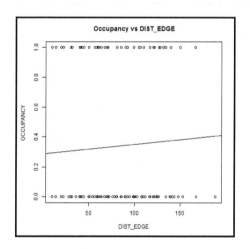

STEP 4: CONDUCT A LINEAR REGRESSION USING MULTIPLE EXPLANATORY VARIABLES USING GENERALISED LINEAR MODELLING (GLM) IN R:

In this step, you will run a second linear regression. This will differ from the one run in step 1 of this exercise because rather than only including elevation as an explanatory variable, it will also include the additional environmental variables linked to the occupancy data in step 2. This can be done by adapting the R code used to create the original GLM to include these additional environmental variables as explanatory variables. To run a GLM using multiple explanatory variables enter the following code into your R WORKSPACE (this is COMMAND 16 in R_CODE_GIS_WORKBOOK_1.DOC):

```
GLM<-glm(OCCUPIED ~ ELEVATION+SLOPE+HILLSHADE+
   DIST_LOCH+DIST_EDGE, family=binomial(link='logit'),
      data= BLUE_TIT_OCCUPANCY_WITH_ALL_EGVS)
                  summary(GLM)
```

Once this command has finished running, the results of this new GLM (which have been called up by the `summary` command) will be displayed in your R WORKSPACE window. They should look like this:

```
Deviance Residuals:
    Min       1Q   Median       3Q      Max
-1.3058  -0.8997  -0.8250   1.4086   1.7791

Coefficients:
             Estimate Std. Error z value Pr(>|z|)
(Intercept) -0.607407   0.946608  -0.642   0.5211
ELEVATION   -0.022195   0.018044  -1.230   0.2187
SLOPE        0.006797   0.025300   0.269   0.7882
HILLSHADE   -0.002358   0.005831  -0.404   0.6859
DIST_LOCH    0.002379   0.001368   1.739   0.0821 .
DIST_EDGE    0.005207   0.004287   1.215   0.2245
---
Signif. codes:  0 '***' 0.001 '**' 0.01 '*' 0.05 '.' 0.1 ' ' 1

(Dispersion parameter for binomial family taken to be 1)

    Null deviance: 252.06  on 197  degrees of freedom
Residual deviance: 247.67  on 192  degrees of freedom
AIC: 259.67

Number of Fisher Scoring iterations: 4

>
```

If you examine the results of this GLM, you will see that none of the explanatory variables are significant at the threshold level of p=0.05 (the p-value for each variable is the last figure on the line that starts with the variable's name). The variable that is closest to being significant is DIST_LOCH, which has a p-value of 0.0821 (highlighted here with a black box, which will not appear on your output). At this stage, this may seem a bit disheartening, but there are still additional analyses that can be conducted to explore the possible relationships with environmental variables in the data set from this study. Specifically, so far, you have only looked at nest box occupancy. However, the original data set also includes measures of breeding success, such as the clutch size, the number of chicks that hatched and the number of chicks that fledged from each nest. In the next exercise, you will investigate whether there are any significant relationships between any of these measures of breeding success and the local environmental variables at the location of each nest box.

Optional extra:

If you wish to get more experience at creating data sets and running linear regression analyses, you can repeat this exercise to look the relationship between nest box occupancy and environmental variables for a second species, the great tit. In order to be able to do this, you will need to have completed the optional extras for exercises one and three in this workbook.

--- Chapter Seven ---

Exercise Five: Conducting Non-Linear Regressions With Generalised Additive Modelling (GAM) Using QGIS And R

So far in the exercises in this workbook, you have used a t-test and linear regression to investigate whether there are any relationships between blue tit nest box occupancy and the local environmental variables within the patch of native oak woodland where the nest boxes are located, but you have yet to find any significant relationships. However, as well as information on nest box occupancy, there is also information available on three measures of breeding success for each occupied nest box. These are the number of eggs laid in each one, the number of chicks that hatched from these eggs, and the number of chicks that fledged. In this exercise, you will investigate whether there are any relationships between these measures of breeding success and the same environmental variables you used to investigate the relationships with occupancy. For these, you will use a different type of analysis. This is generalised additive modelling (GAM). GAMs can be used to investigate whether any non-linear relationships exist in your data, and the main advantage over GLMs is the fact that the shape of the relationship is not decided before running the model, but is, instead, determined by the modelling process itself.

In order to be able to run a GAM in R, you first need to load the MGCV packaged into R (**NOTE:** You will only need to do this once). To do this, enter the following line of code into your R WORKSPACE window:

```
install.packages("mgcv")
```

Exercise Five: Conducting Non-Linear Regressions With GAMs Using QGIS And R

Once this code has been entered, follow any instructions that appear to download and install the required package in your version of R. After it has been installed, you will be able to access the tools required to run the GAMs which you will do in this exercise

To investigate whether there are any relationships between breeding success and the location where a nest box is sited, you will create GAMs that will compare each of the three measures of breeding success (clutch size, number of hatchlings and number of fledglings recorded in each nest box occupied by blue tits) to all the local environmental variables for which you generated raster data layers in exercise two. This will be done using R in step 3 of this exercise.

However, before you can create your GAMs, you need to process your existing nest box data in a number of different ways. Firstly, you need to create a subset which only includes the nest boxes occupied by blue tits during the breeding season under investigation. This is because you can only examine relationships with these measures of breeding success in nest boxes where breeding actually took place. Secondly, you will need to link the information on breeding success to the data set containing just these nest boxes. This data processing will be done in QGIS in steps 1 and 2 of this exercise.

The starting point for this exercise is the GIS project created in exercise four. Open QGIS. Once it is open, click on the PROJECT menu and select OPEN. When the OPEN window appears, browse to the location where your project is saved (C:\QGIS_R_WORKBOOK), select it and click OPEN. After you have opened the GIS project called EXERCISE_FOUR, the first thing you need to do is save it under a new name. This is because you do not want to alter the contents of the original project, you just want to base your new one on it since this saves you having to add all the data layers again, and also having to reset the projection/coordinate system of the data frame. To save the project under a new name, click on PROJECT on the main menu bar, and select SAVE AS. For this exercise, save it as EXERCISE_FIVE in the C:\QGIS_R_WORKBOOK folder.

Before you start this exercise, move the table called NEST_BOX_BREEDING_DATA to the top of your TABLE OF CONTENTS window by clicking on its name and dragging into the desired position. Repeat this process for the data layer called BLUE_TIT_

Exercise Five: Conducting Non-Linear Regressions With GAMs Using QGIS And R

OCCUPANCY_WITH_ALL_EGVS (created in exercise four). Now turn off all the raster data layers by clicking on the boxes next to their names in the TABLE OF CONTENTS window so that are no longer displayed in the MAP window. Do the same for the data layers called SCENE_NESTBOX_LOCATIONS and BLUE_TIT_OCCUPANCY_ WITH_ELEVATION. Next, right click on the data layer called SCENE_NESTBOX_ LOCATIONS and select STYLES> COPY STYLE. Now right click on the data layer called BLUE_TIT_OCCUPANCY_WITH_ALL_EGVS and select STYLES> PASTE. At this point you may notice that the symbols for this layer disappear. This is okay and you will sort this out next. Right click on the name of the data layer called BLUE_TIT_ OCCUPANCY_WTH_ALL_EGVS in the TABLE OF CONTENTS window and select PROPERTIES. In the LAYER PROPERTIES window that opens, click on the STYLE tab and then for COLUMN, select OCCUPIED before clicking the OK button to close this window. The symbols for this layer should now re-appear in the MAP window. At this point, your TABLE OF CONTENTS window should look like this

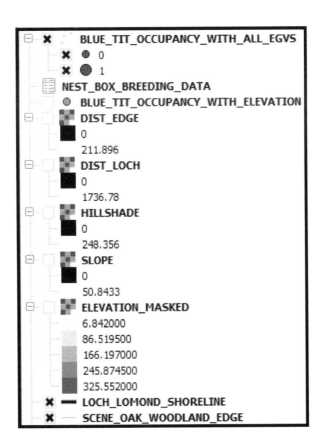

While the contents of your MAP window should look like this

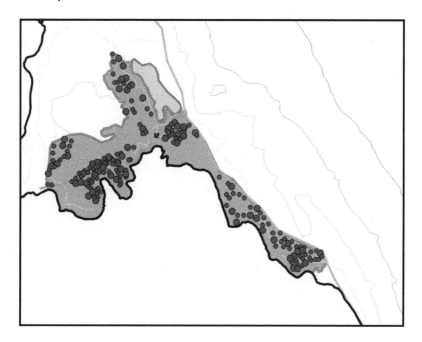

STEP 1: CREATE A NEW POINT DATA LAYER IN QGIS THAT ONLY CONTAINS THE NEST BOXES OCCUPIED BY BLUE TITS:

To create a data layer which contains only the nest boxes occupied by blue tits, you need to first select these boxes and then create a new data layer from them. This can be done using the SELECT BY EXPRESSION tool to select the desired points based on the contents of the OCCUPIED field. Once these points have been selected, you can use the SAVE AS tool to make a new data layer that contains only these selected features. To do this, work through the flow diagram on the next page.

Exercise Five: Conducting Non-Linear Regressions With GAMs Using QGIS And R

Point data layer from which you wish to create a new data layer containing a subset of its features

The point data layer from which you will create a new data layer containing a subset of its features in this step is BLUE_TIT_OCCUPANCY_WITH_ALL_EGVS.

Click on the name of the data layer BLUE_TIT_OCCUPANCY_WITH_ALL_EGVS in the TABLE OF CONTENTS window so that it is highlighted. Next, click on VIEW on the main menu bar, and select SELECT> SELECT BY EXPRESSION. In the SELECT BY EXPRESSION window that opens, enter the expression:

"OCCUPIED" = 1

Now click on the SELECT button. This will select all the nest boxes which were occupied by blue tits in the breeding season being examined. Now click CLOSE to close the SELECT BY EXPRESSION window. You can examine the selection you just made in the MAP window to ensure that only the nest boxes occupied by blue tits have been select (these will be the larger symbols).

1. Make a new data layer by selecting a subset of data in an existing data layer

To make a new data layer containing only the selected nest boxes, right click on the data layer called BLUE_TIT_OCCUPANCY_WITH_ALL_EGVS in the TABLE OF CONTENTS window and select SAVE AS. This opens the SAVE VECTOR LAYER AS window. In this window, for FORMAT, select ESRI SHAPEFILE. Next, enter the following address and file name into the SAVE AS section of the window: C:/QGIS_R_WORKBOOK/BLUE_TIT_BREEDING_DATA_WITH_ALL_EGVS. In the CRS section, make sure that the option starting with LAYER CRS is selected. Now, select SAVE ONLY SELECTED FEATURES and then select ADD SAVED FILE TO MAP. Finally, to save the selected records as a new data layer, click on the OK button.

2. Select how your new point data layer will be displayed

To copy the style from the BLUE_TIT_OCCUPANCY_WITH_ALL_EGVS, right click on the name of this data layer in the TABLE OF CONTENTS window and select STYLES> COPY STYLE. Next, right click on the name of the data layer created in stage 1 (BLUE_TIT_BREEDING_DATA_WITH_ALL_EGVS) and select STYLES> PASTE STYLE. Finally, click OK to close the LAYER PROPERTIES window.

New point data layer created containing a subset of the original features

Once you have completed this step, your TABLE OF CONTENTS window should look like this:

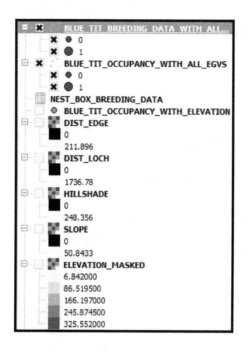

While the contents of your MAP window should look like this:

You can now check that the new data layer created in this step (BLUE_TIT_BREEDING_DATA_WITH_ALL_EGVS) has the required features in it by clicking on the box next to the data layer called BLUE_TIT_OCCUPANCY_WITH_ALL_EGVS so that it is no longer displayed in the MAP window. The contents of the MAP window should now look like this:

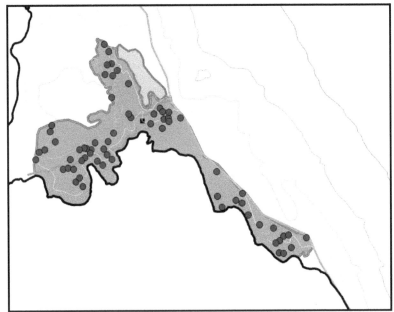

This ability to do a quick and easily visual check on the contents of any new data subsets to ensure that it contains the correct records is one of the main advantages of doing this step in QGIS rather than in R.

STEP 2: LINK THE DATA ON BREEDING SUCCESS TO THE NEST BOXES OCCUPIED BY BLUE TITS IN QGIS AND PREPARE THE COMBINED DATA SET FOR ANALYSIS:

Now that you have created a data layer that only contains the nest boxes occupied by blue tits during one particular breeding season, you need to join the information on breeding success to it. This information is contained in the table NEST_BOX_BREEDING_DATA added to your GIS project in step 3 of exercise one. To join the required data from this table to the attribute table of the data layer created in step 1 of this exercise, work through the flow diagram provided on the next page.

Exercise Five: Conducting Non-Linear Regressions With GAMs Using QGIS And R

Point data layer of nest box locations and table with information on breeding success

The point data layer of the nest box locations is called BLUE_TIT_BREEDING_DATA_WITH_ALL_EGVS, while the table with the information about breeding success is called NEST_BOX_BREEDING_DATA.

1. Join the breeding success data to the attribute table of the nest box locations point data layer

Right click on the data layer called BLUE_TIT_BREEDING_DATA_WITH_ALL_EGVS in the TABLE OF CONTENTS and select PROPERTIES. In the LAYER PROPERTIES window, click on the JOINS tab and then click on the ADD JOIN button (it is at the bottom of the tab and has a green cross on it). In the ADD VECTOR JOIN window that opens, select NEST_BOX_BREEDING_DATA as the JOIN LAYER, and select NEST_BOX as the JOIN FIELD. For TARGET FIELD, select BOX_NUMBER. This will join the data from the NEST_BOX_BREEDING_DATA table to the attribute table for BLUE_TIT_BREEDING_DATA_WITH_ALL_EGVS data layer based on the contents of these fields. Next, click on the box next to CHOOSE WHICH FIELDS ARE JOINED and select the fields called CLUTCH SIZE, NO CHICKS HATCHED and NO CHICKS END. These fields contain the information about these measures of breeding success for each nest box during the breeding season being examined. Now click on the box next to CUSTOM FIELD NAME PREFIX, and enter BD_ (which stands for Breeding Data) in this section. Finally, click OK to make the join, and then click OK to close the LAYER PROPERTIES window.

Table joins in QGIS are only temporary. This means that before you can analyse the data on breeding success, you need to create a permanent field for each measurement and transfer the required data into it. To do this, first right click on the data layer called BLUE_TIT_BREEDING_DATA_WITH_ALL_EGVS in the TABLE OF CONTENTS window and select OPEN ATTRIBUTE TABLE. In the TABLE window that opens, click on the TOGGLE EDITING button in the top left hand corner (it has a picture of a pencil on it), and then click on the OPEN FIELD CALCULATOR button at the top of the TABLE window (it as a picture of an abacus on it). In the window that opens, select CREATE NEW FIELD and then for OUTPUT FIELD NAME, enter NO_EGGS (making sure it is typed exactly as written here, in uppercase letters and with an underscore). Once you have done this, select WHOLE NUMBER (INTEGER) for OUTPUT FIELD TYPE. Next, click on FIELDS AND VALUES under FUNCTION and then double-click on BD_CLUTCH SIZE to add it to the EXPRESSION section. Now click OK to create this field and transfer the data from the BD_CLUTCH SIZE field into a new permanent field called NO_EGGS.

2. Create a permanent field for the data from the BD_CLUTCH SIZE field and transfer these data into it from this temporary join field.

Exercise Five: Conducting Non-Linear Regressions With GAMs Using QGIS And R

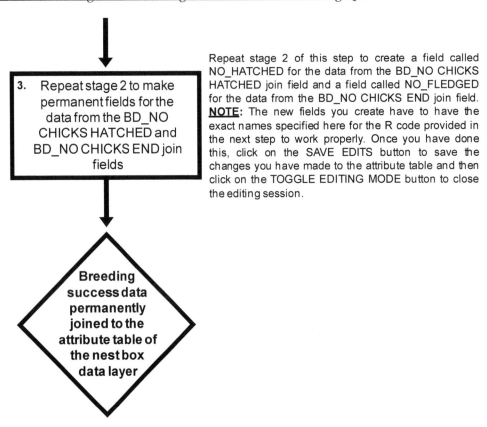

You can now carefully examine the contents of the different fields of the attribute table of the BLUE_TIT_BREEDING_DATA_WITH_ALL_EGVS data layer to ensure that the data have been correctly joined together. Again, this ability to quickly and easily examine the contents of your joined data set is one of the main advantages of doing this step in QGIS rather than in R. Once you are happy that your data have been correctly joined together, you can remove the temporary join field from your attribute table. To do this, first close the TABLE window. Next, right click on the name of the BLUE_TIT_BREEDING_ DATA_WITH_ALL_EGVS in the TABLE OF CONTENTS window and select PROPERTIES. In the LAYER PROPERTIES window that opens, click on the JOINS tab. In the JOINS tab, select the join called NEST_BOX_BREEDING_DATA and then click on the button with the red minus sign (-) on it at the bottom of the window to remove the join. You can now click OK to close the LAYER PROPERTIES window. Finally, right click on the name of the BLUE_TIT_BREEDING_DATA_WITH_ALL_EGVS in the TABLE OF CONTENTS window and select OPEN ATTRIBUTE TABLE.

The contents of the TABLE window should now look like this:

	BOX_NUMBER	LATITUDE	LONGITUDE	OCCUPIED	ELEVATION	EL_CAT	SLOPE	HILLSHADE	DIST_LOCH	DIST_EDGE	NO_EGGS	NO_HATCHED	NO_FLEDGED
0	142	56.12621599999...	-4.61699800000...	1	20.00000	10-20	7.80331	155.47945	44.72136	50.00000	6	6	6
1	24	56.13129000000...	-4.61525100000...	1	22.87749	20-30	4.61178	112.82821	364.00549	70.00000	12	12	12
2	48	56.12906499999...	-4.61481100000...	1	25.00000	20-30	22.64382	113.54879	134.53624	128.06248	12	9	8
3	301	56.13180899999...	-4.61724900000...	1	32.92912	30-40	14.13227	117.36760	470.10638	80.62257	10	8	NULL
4	44	56.13033200000...	-4.61723800000...	1	60.00000	50 or more	26.76279	35.79152	336.15472	14.14214	9	9	9
5	3	56.12926999999...	-4.61511000000...	1	32.50000	30-40	15.22593	148.58337	164.01219	120.83046	10	8	8
6	300	56.13213900000...	-4.61671900000...	1	25.14756	20-30	9.02904	114.15614	488.46698	92.19544	6	7	7
7	308	56.13259999999...	-4.61738300000...	1	25.85823	20-30	8.83440	105.29299	550.36353	70.71068	7	7	7
8	309	56.13340500000...	-4.61780600000...	1	22.50000	20-30	12.08868	140.13589	640.78076	30.00000	10	10	9
9	315	56.13387999999...	-4.61841900000...	1	25.66380	20-30	10.53481	97.97186	592.28369	10.00000	9	9	9
10	304	56.13184499999...	-4.61821900000...	1	41.11111	40-50	5.77516	124.36996	510.39200	40.00000	8	8	4
11	307	56.13252299999...	-4.61793500000...	1	33.33333	30-40	18.46657	116.69751	559.46405	92.19544	8	8	8
12	73	56.12795400000...	-4.61687300000...	1	46.66667	40-50	20.92531	142.66203	120.83046	120.83046	10	9	9
13	94	56.12737099999...	-4.61756200000...	1	40.00000	30-40	24.61110	203.75623	98.99495	98.99495	7	7	0
14	108	56.12725900000...	-4.61914900000...	1	43.33333	40-50	21.01417	196.32610	162.78821	120.83046	6	6	6
15	99	56.12688800000...	-4.61808200000...	1	26.66667	20-30	7.60489	153.20044	104.40307	104.40307	7	7	7
16	112	56.12680100000...	-4.61960300000...	1	25.00000	20-30	15.82082	180.21437	110.00000	110.00000	6	6	6
17	114	56.12685499999...	-4.62026100000...	1	30.00000	20-30	22.61987	202.62941	120.41595	121.65525	7	7	7
18	118	56.12688599999...	-4.62050400000...	1	30.00000	20-30	20.78340	197.48854	123.69317	126.49110	8	6	4
19	191	56.12656799999...	-4.61974900000...	1	20.00000	10-20	17.39023	187.52913	90.00000	90.00000	7	7	5
20	118	56.12677200000...	-4.62112000000...	1	36.66667	30-40	28.72484	179.24379	130.38405	134.16408	8	6	4

STEP 3: CONDUCT GAMS USING EACH OF THE THREE MEASURES OF BREEDING SUCCESS AS THE RESPONSE VARIABLE USING R:

Now that you have successfully created a data set that contains just the nest boxes used by blue tits during a particular breeding season, along with information about the local environmental variables for each nest box and three measures of breeding success, you are ready to create your GAMs to test whether there are any significant relationships between blue tit breeding success and these environmental variables. However, before you can do this, you need to import the new data set incorporating the measures of breeding success for blue tits created in step 2 of this exercise into R. This can be done by selecting the contents of the attribute table of the data layer called BLUE_TIT_BREEDING_DATA_WITH_ALL_EGVS in QGIS (which should still be open at this point) and copying it to the clipboard of your operating system. It can then be imported into R by running the following command:

```
BLUE_TIT_BREEDING_DATA_WITH_ALL_EGVS <-read.table(file =
        "clipboard", sep = "\t", header=TRUE)
```

Exercise Five: Conducting Non-Linear Regressions With GAMs Using QGIS And R

This code has to be entered exactly as it is written here or it will not work. If you wish to use the copy-and-paste approach for entering the R commands, copy the block of text that is directly below COMMAND 17 in the document R_CODE_GIS_WORKBOOK_1.DOC. **NOTE:** If you are using a computer with a Mac OS, you will need to replace the text `"clipboard"` in this command with the text `pipe("pbpaste")` in order for it to work.

This command will create a new object in R called BLUE_TIT_BREEDING_DATA_WITH_ALL_EGVS that contains the information on the three measures of breeding success as well as the information about the local environmental variables at the location of each nest box occupied by blue tits. If, when your run this `read.table` command, you get an error message containing the text *"Incomplete final line found ..."*, return to QGIS, copy the contents of the attribute table to your clipboard again, and then re-run this command without copy-and-pasting it into your R Workspace again. One way to do this is to use the UP ARROW key on your keyboard to bring the last piece of code you ran back on to the command line in your R WORKSPACE window before pressing the ENTER key to run it again.

Once you have successfully create the new R object containing the information on breeding success within the nest boxes occupied by blue tits, you can create your first GAM by working through the flow diagram provided on the next page.

NOTE: This GAM will be run for the sole purpose of providing experience with integrating QGIS and R. As a result, it does not include any of the exploration steps that you should conduct on your data before running a GAM, or any of the steps required to ensure that there are no problems with it (such as looking for patterns in residuals or testing for over-dispersion). If you are doing you own analysis, you need to ensure that you include these steps. More information about what these additional steps are can be found in most statistical textbooks, such as those by Alain Zuur (see *www.highstat.com/books.htm* for details).

Exercise Five: Conducting Non-Linear Regressions With GAMs Using QGIS And R

Data set of nest box locations with information about breeding success and environmental variables

For this exercise, this data set will be the R object called BLUE_TIT_BREEDING_DATA_WITH_ALL_EGVS created at the start of this step.

To run a GAM on the relationship between the clutch size of blue tits in occupied nest boxes and the local environmental variables, enter the following code into your R WORKSPACE window:

```
library(mgcv)
GAM<-gam(NO_EGGS~+s(ELEVATION,fx=F,k=4)
    +s(SLOPE,fx=F,k=4)
    +s(HILLSHADE,fx=F,k=4)
    +s(DIST_LOCH,fx=F,k=4)
    +s(DIST_EDGE,fx=F,k=4),
    family=poisson, data=BLUE_TIT_
    BREEDING_DATA_WITH_ALL_EGVS)
summary(GAM)
plot(GAM, pages=1)
```

1. Run a GAM on the relationship between clutch size and local environmental variables

This code has to be entered exactly as it is written here or it will not work. If you wish to use the copy-and-paste approach for entering the R commands, copy the block of text that is directly below COMMAND 18 in the document R_CODE_GIS_WORKBOOK_1.DOC.

This code uses the `gam` command to create a GAM based on the information in the fields ELEVATION, SLOPE, HILLSHADE, DIST_LOCH, DIST_EDGE and NO_EGGS in the dataset BLUE_TIT_BREEDING_DATA_WITH_ALL_EGVS. Within the `gam` code, the `family=poisson` term is used to denote that the regression being run should be based on a poisson distribution rather than one based on any other possible distribution that can be used with the `gam` function. Finally, the `summary` command is used to call up and display the results of this analysis, and the `plot` command is used to call up and display the GAM smoothers that show the relationships between the explanatory variables and the response variable (in this case, the number of eggs laid in each nest box).

GAM conducted using clutch size as the response variable

111

Exercise Five: Conducting Non-Linear Regressions With GAMs Using QGIS And R

Once this command has finished running, the R WORKSPACE window should be displaying the results of your first GAM (which has been called up by the `summary` command), and they should look like this:

```
Family: poisson
Link function: log

Formula:
NO_EGGS ~ +s(ELEVATION, fx = F, k = 4) + s(SLOPE, fx = F, k = 4) +
    s(HILLSHADE, fx = F, k = 4) + s(DIST_LOCH, fx = F, k = 4) +
    s(DIST_EDGE, fx = F, k = 4)

Parametric coefficients:
            Estimate Std. Error z value Pr(>|z|)
(Intercept)  2.08474    0.04345   47.98   <2e-16 ***
---
Signif. codes:  0 '***' 0.001 '**' 0.01 '*' 0.05 '.' 0.1 ' ' 1

Approximate significance of smooth terms:
             edf Ref.df Chi.sq p-value
s(ELEVATION)   1      1  0.134   0.715
s(SLOPE)       1      1  0.110   0.740
s(HILLSHADE)   1      1  1.411   0.235
s(DIST_LOCH)   1      1  0.264   0.607
s(DIST_EDGE)   1      1  0.384   0.535

R-sq.(adj) =  -0.0246   Deviance explained = 4.13%
UBRE = 0.0337  Scale est. = 1         n = 66
> plot(GAM, pages=1)
```

If you examine this GAM summary, you will see that none of the environmental variables included in the model had a significant effect on the clutch size of blue tits in these nest boxes (the p-values are provided at the end of the line of figures after each variable's name).

You can now repeat this step using the number of hatchlings in each nest box as the measure of breeding success. This information is contained in the field called NO_HATCHED in the R object called BLUE_TIT_BREEDING_DATA_WITH_ ALL_EGVS created at the start of this step. To do this, replace NO_EGGS with NO_HATCHED within the R code for the GAM provided in the above flow diagram (see COMMAND 19 in R_CODE_GIS_WORKBOOK_1.DOC).

Once this command has finished running, the R WORKSPACE window should be displaying the results of your second GAM using the number of hatchlings per nest box as the response variable, and it should look like the image at the top of the next page.

```
Family: poisson
Link function: log

Formula:
NO_HATCHED ~ +s(ELEVATION, fx = F, k = 4) + s(SLOPE, fx = F,
    k = 4) + s(HILLSHADE, fx = F, k = 4) + s(DIST_LOCH, fx = F,
    k = 4) + s(DIST_EDGE, fx = F, k = 4)

Parametric coefficients:
            Estimate Std. Error z value Pr(>|z|)
(Intercept)   2.0380     0.0463   44.02   <2e-16 ***
---
Signif. codes:  0 '***' 0.001 '**' 0.01 '*' 0.05 '.' 0.1 ' ' 1

Approximate significance of smooth terms:
              edf Ref.df Chi.sq p-value
s(ELEVATION) 1.000  1.000  0.149   0.700
s(SLOPE)     1.000  1.000  0.290   0.590
s(HILLSHADE) 1.000  1.000  0.441   0.507
s(DIST_LOCH) 1.086  1.168  0.440   0.535
s(DIST_EDGE) 1.000  1.000  0.172   0.679

R-sq.(adj) =  0.0499   Deviance explained = 13%
UBRE = -0.37916  Scale est. = 1         n = 61
> plot(GAM, pages=1)
```

Again, if you examine this GAM summary, you will see that none of the environmental variables included in the model had a significant effect on the number of blue tit hatchlings per nest box.

Finally, repeat this step using the number of fledglings in each nest box as the measure of breeding success. This information is contained in the field called NO_FLEDGED in the R object called BLUE_TIT_BREEDING_DATA_WITH_ALL_EGVS created at the start of this step. To do this, replace NO_EGGS with NO_FLEDGED within the R code for the GAM provided in the above flow diagram (see COMMAND 20 in R_CODE_GIS_WORKBOOK_1.DOC).

Once this command has finished running, the R WORKSPACE window should be displaying the results of your third GAM using the number of fledglings per nest box as the response variable, and it should look like the image at the top of the next page.

Exercise Five: Conducting Non-Linear Regressions With GAMs Using QGIS And R

```
Family: poisson
Link function: log

Formula:
NO_FLEDGED ~ +s(ELEVATION, fx = F, k = 4) + s(SLOPE, fx = F,
    k = 4) + s(HILLSHADE, fx = F, k = 4) + s(DIST_LOCH, fx = F,
    k = 4) + s(DIST_EDGE, fx = F, k = 4)

Parametric coefficients:
             Estimate Std. Error z value Pr(>|z|)
(Intercept)   1.79665    0.05694   31.55   <2e-16 ***
---
Signif. codes:  0 '***' 0.001 '**' 0.01 '*' 0.05 '.' 0.1 ' ' 1

Approximate significance of smooth terms:
                edf Ref.df Chi.sq  p-value
s(ELEVATION)  1.000  1.000  3.336 0.067785 .
s(SLOPE)      2.075  2.491  6.888 0.049728 *
s(HILLSHADE)  1.000  1.000  5.633 0.017627 *
s(DIST_LOCH)  2.753  2.956 18.884 0.000254 ***
s(DIST_EDGE)  1.000  1.001  0.556 0.456258
---
Signif. codes:  0 '***' 0.001 '**' 0.01 '*' 0.05 '.' 0.1 ' ' 1

R-sq.(adj) =  0.237   Deviance explained = 27.1%
UBRE = 1.0941  Scale est. = 1         n = 53
> plot(GAM, pages=1)
```

If you examine this GAM summary, you will see that it has identified three significant relationships. These are the relationships with the slope of the local terrain (SLOPE – p=0.049728), the amount of direct sunlight the location of each nest box receives at midday in early summer (HILLSHADE – p=0.017627) and the distance to the shoreline of Loch Lomond (DIST_LOCH – p=0.000254). To find out more about the form of these relationships, you can examine the graphs called up by the plot command at the end of the command used to generate this GAM. These will be displayed in an R GRAPHICS window, and its contents should look like this:

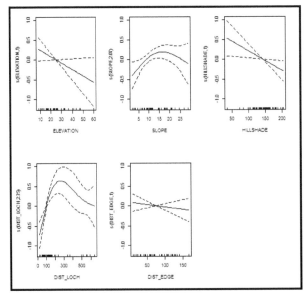

These are partial plots of the GAM smoothers for this model and they give an indication of the shape of the relationships between the explanatory variables and the response variable. These can be identified by relative positions of the 95% confidence intervals (dotted lines) around the best fit trend line (solid line) on the partial plots. The partial plot for the SLOPE smoother indicates that there is a generally positive relationship between the slope of the local terrain and the number of blue tit fledglings in each nest box until it reaches a value of about 10 degrees, where this relationship disappears. The partial plot for the HILLSHADE smoother indicates that there is a negative linear relationship between the amount of direct sunlight the location where a nest box is sited gets in at midday in early summer and the number of fledglings. Finally, the partial plot for the DIST_LOCH smoother indicates a positive relationship between the distance from the shoreline of Loch Lomond and the number of fledglings until a distance of approximately 150 meters, at which point this relationship disappears. The partial plots for the remaining smoothers can be ignored as the relationships with these variables were not found to be significant. More detailed information on interpreting the relationships identified from the partial plots of GAM smoothers can be found in *A Beginner's Guide To Generalized Additive Models With R* by Alain Zuur.

At the moment, you have simply included all possible response variables within the model. However, in order to refine it, you may decide you wish to undertake model selection to identify the best model from all possible combinations of the five variables included in it. You can do this editing the R code for this final GAM (see COMMAND 20 in R_CODE_GIS_WORKBOOK_1.DOC) to include different combinations of variables. You may also need to add some additional code to calculate a metric, such as the Akaike Information Criteria (AIC), which you can use to objectively determine which is the best model. To calculate an AIC value, you can add the following command to the end of the block of code you used to create the GAM in the flow diagram in page 111:

```
AIC(GAM)
```

The R code required to provide other metrics that can be used during model selection can be found using an internet search engine by entering the name of the metric you wish to calculate followed by the term *R Code*.

Optional extra:

If you wish to get further experience in running a GAM on breeding success using QGIS and R, you can also test whether there are any significant relationships between environmental variables and other measures of breeding behaviour, including the date that the clutch of eggs in each occupied box was completed (COMPLETION DATE) and the date that the eggs hatched (OBSERVED HATCHDATE). Similarly, you can repeat this exercise using great tits as the target species. However, there is only one measure of breeding success (clutch size) provided in the NEST_BOX_BREEDING_DATA table for this species.

Appendix I:
List Of Commands Containing The R Code Used Within This Workbook

This appendix contains all the R code required to complete the exercises in this workbook. These same commands are also provided in the document R_CODE_GIS_WORKBOOK_1.DOC that is included in the compressed folder of data files downloaded at the start of exercise one. However, the commands in the R_CODE_GIS_WORKBOOK_1.DOC file are colour-coded to help you develop a greater understanding of exactly what each piece of code does. This code is provided on an 'as is' basis and is provided for training purposes only. If it is used for any other purpose, it is the responsibility of the end user to ensure this code is accurate and appropriate for the specific purpose for which it is to be used. Neither the authors or the publisher will be held responsible for any errors within this code. It has only been tested with R version 3.6.1 and there are no guarantees it will work with other versions of R, but there is no reason that it shouldn't.

NOTE: Whenever you need to transfer your data from the attribute table of a data layer in QGIS to R, open the attribute table in by right-clicking on the name of the data layer in the TABLE OF CONTENTS window. Next, select the entire contents of the attribute table by clicking on the grey box just above 0 on the left hand side, and then press CTRL+C on your keyboard. This will copy the contents of the attribute table to your clipboard. To import it into R, use the following command:

```
read.table(file = "clipboard", sep = "\t", header=TRUE)
```

If you are using a Mac OS computer, you will need to replace the text `"clipboard"` with the text `pipe("pbpaste")` for this command to work. This applies throughout the R code provided below.

Appendix I: R Code Used In The Exercises In This Workbook

NOTE: The commands below assume you have exactly followed the naming protocols provided in the instructions for completing each step in the exercises in this workbook. This includes the same usage of uppercase and lowercase letters. If you have used different names for any data layers, field names, field contents or objects created in R, you will have to adapt these commands accordingly.

COMMAND 1

(This command imports the contents of an attribute table that has been copied to the clipboard of your operating system.)

Before you run this command in R, open the attribute table of the data layer called BLUE_TIT_OCCUPANCY_WITH_ELEVATION in QGIS and copy its contents to your clipboard.

```
BLUE_TIT_OCCUPANCY_WITH_ELEVATION<-read.table(file =
      "clipboard", sep = "\t", header=TRUE)
```

This code creates an object in R called BLUE_TIT_OCCUPANCY_WITH_ELEVATION which contains the contents of attribute table which you copied from QGIS. This is the name the code below assumes has been used for this data set in R. **NOTE:** If you are using a Mac OS computer, you will need to replace the text `"clipboard"` with the text `pipe("pbpaste")` for this command to work.

COMMAND 2

(This command creates a histogram of the distribution of all nest boxes in the selected data layer in relation to elevation.)

```
hist(BLUE_TIT_OCCUPANCY_WITH_ELEVATION$ELEVATION,nclass=8)
```

COMMAND 3

(This command creates a histogram of the distribution of just the occupied nest boxes in the selected data layer in relation to elevation.)

```
OCCUPIED_BOXES=subset(BLUE_TIT_OCCUPANCY_WITH_ELEVATION,
                  OCCUPIED=="1")
      hist(OCCUPIED_BOXES$ELEVATION,nclass=8)
```

Appendix I: R Code Used In The Exercises In This Workbook

COMMAND 4
(This command calculates the occupancy rate of nest boxes in each elevation category and then creates a bar graph from these data.)

```
NESTBOX_OCCUPANCY=
aggregate(BLUE_TIT_OCCUPANCY_WITH_ELEVATION$OCCUPIED,
list(BLUE_TIT_OCCUPANCY_WITH_ELEVATION$EL_CAT),sum)
colnames(NESTBOX_OCCUPANCY)=c("EL_CAT","OCCUPIED")
NESTBOX_OCCUPANCY$N=
table(BLUE_TIT_OCCUPANCY_WITH_ELEVATION$EL_CAT)
NESTBOX_OCCUPANCY$OCCUPANCY=((NESTBOX_OCCUPANCY$OCCUPIED/
NESTBOX_OCCUPANCY$N)*100)
barplot(NESTBOX_OCCUPANCY$OCCUPANCY,xlab="ELEVATION
CATEGORIES", ylab="OCCUPANCY RATES",
names.arg=NESTBOX_OCCUPANCY$EL_CAT)
```

COMMAND 5
(This command calculates the mean elevation for occupied and unoccupied nest boxes.)

```
OCCUPIED_BOXES=subset(BLUE_TIT_OCCUPANCY_WITH_ELEVATION,
OCCUPIED=="1")
mean(OCCUPIED_BOXES$ELEVATION)
UNOCCUPIED_BOXES=subset(BLUE_TIT_OCCUPANCY_WITH_ELEVATION,
OCCUPIED=="0")
mean(UNOCCUPIED_BOXES$ELEVATION)
```

COMMAND 6
(This command runs a t-test to compare the mean elevations of occupied and unoccupied nest boxes.)

```
t.test(ELEVATION ~ OCCUPIED, data =
BLUE_TIT_OCCUPANCY_WITH_ELEVATION, var.equal = TRUE)
```

Appendix I: R Code Used In The Exercises In This Workbook

COMMAND 7
(This command creates a table of summary statistics for the elevation values of nest boxes occupied by blue tits.)

```
OCCUPIED_BOXES=subset(BLUE_TIT_OCCUPANCY_WITH_ELEVATION,
                     OCCUPIED=="1")
SUMMARY_STATISTICS_1<-
  data.frame(rbind(length(OCCUPIED_BOXES$ELEVATION),
    length(unique(OCCUPIED_BOXES$ELEVATION)),
        min(OCCUPIED_BOXES$ELEVATION),
        max(OCCUPIED_BOXES$ELEVATION),
        max(OCCUPIED_BOXES$ELEVATION)-
        min(OCCUPIED_BOXES$ELEVATION),
        mean(OCCUPIED_BOXES$ELEVATION),
        median(OCCUPIED_BOXES$ELEVATION),
        sd(OCCUPIED_BOXES$ELEVATION)),
   row.names=c("Count:","Unique values:","Minimum
value:","Maximum value:","Range:","Mean value:","Median
         value:","Standard deviation:"))
colnames(SUMMARY_STATISTICS_1)<-c("ELEVATION OF OCCUPIED
                       BOXES")
SUMMARY_STATISTICS_1
```

COMMAND 8
(This command creates a table of summary statistics for the elevation values of the nest boxes not occupied by blue tits.)

```
UNOCCUPIED_BOXES=subset(BLUE_TIT_OCCUPANCY_WITH_ELEVATION,
                     OCCUPIED=="0")
SUMMARY_STATISTICS_2<-
  data.frame(rbind(length(UNOCCUPIED_BOXES$ELEVATION),
    length(unique(UNOCCUPIED_BOXES$ELEVATION)),
        min(UNOCCUPIED_BOXES$ELEVATION),
        max(UNOCCUPIED_BOXES$ELEVATION),
        max(UNOCCUPIED_BOXES$ELEVATION)-
        min(UNOCCUPIED_BOXES$ELEVATION),
        mean(UNOCCUPIED_BOXES$ELEVATION),
        median(UNOCCUPIED_BOXES$ELEVATION),
        sd(UNOCCUPIED_BOXES$ELEVATION)),
   row.names=c("Count:","Unique values:","Minimum
```

```
value:","Maximum value:","Range:","Mean value:","Median
       value:","Standard deviation:"))
colnames(SUMMARY_STATISTICS_2)<-c("ELEVATION OF UNOCCUPIED
                       BOXES")
            SUMMARY_STATISTICS_2
```

COMMAND 9

(This command creates a scatter plot of elevation vs occupancy and adds a line of best fit.)

```
       plot(BLUE_TIT_OCCUPANCY_WITH_ELEVATION$ELEVATION,
BLUE_TIT_OCCUPANCY_WITH_ELEVATION$OCCUPIED, main="OCCUPANCY
   VS ELEVATION", xlab="ELEVATION",  ylab="OCCUPANCY ")
            abline(lm(OCCUPIED~ELEVATION, data=
              BLUE_TIT_OCCUPANCY_WITH_ELEVATION))
```

COMMAND 10

(This command runs a logistic linear regression using a GLM to test whether the linear relationship between nest box occupancy and elevation is significant.)

```
       LOGISTIC_REGRESSION<-glm(OCCUPIED ~ ELEVATION,
            family=binomial(link='logit'),data=
              BLUE_TIT_OCCUPANCY_WITH_ELEVATION)
                summary(LOGISTIC_REGRESSION)
```

COMMAND 11

(This command imports the contents of an attribute table that has been copied to the clipboard of your operating system.)

Before you run this command in R, open the attribute table of the data layer called BLUE_TIT_OCCUPANCY_WITH_ALL_EVGS in QGIS and copy its contents to your clipboard.

```
     BLUE_TIT_OCCUPANCY_WITH_ALL_EGVS<-read.table(file =
          "clipboard", sep = "\t", header=TRUE)
```

This code creates an object in R called BLUE_TIT_OCCUPANCY_WITH_ALL_EGVS which contains the contents of attribute table which you copied from QGIS. This is the name the code below assumes has been used for this data set in R. **NOTE**: If you are using a Mac OS computer, you will need to replace the text "clipboard" with the text pipe("pbpaste") for this command to work.

Appendix I: R Code Used In The Exercises In This Workbook

COMMAND 12
(This command creates a scatter plot of slope vs occupancy and add a best fit line to it.)

```
plot(BLUE_TIT_OCCUPANCY_WITH_ALL_EGVS$SLOPE,
BLUE_TIT_OCCUPANCY_WITH_ALL_EGVS$OCCUPIED, main="OCCUPANCY
     VS SLOPE", xlab="SLOPE",
          ylab="OCCUPANCY")
     abline(lm(OCCUPIED~SLOPE, data=
     BLUE_TIT_OCCUPANCY_WITH_ALL_EGVS))
```

COMMAND 13
(This command creates a scatter plot of hillshade vs occupancy and adds a line of best fit.)

```
plot(BLUE_TIT_OCCUPANCY_WITH_ALL_EGVS$HILLSHADE,
BLUE_TIT_OCCUPANCY_WITH_ALL_EGVS$OCCUPIED, main="OCCUPANCY
  VS HILLSHADE", xlab="HILLSHADE",  ylab="OCCUPANCY")
       abline(lm(OCCUPIED~HILLSHADE, data=
       BLUE_TIT_OCCUPANCY_WITH_ALL_EGVS))
```

COMMAND 14
(This command creates a scatter plot of the distance to the shoreline of Loch Lomond vs occupancy and adds a line of best fit.)

```
plot(BLUE_TIT_OCCUPANCY_WITH_ALL_EGVS$DIST_LOCH,
BLUE_TIT_OCCUPANCY_WITH_ALL_EGVS$OCCUPIED, main="Occupancy
  vs DIST_LOCH", xlab="DIST_LOCH",  ylab="OCCUPANCY")
       abline(lm(OCCUPIED~DIST_LOCH, data=
       BLUE_TIT_OCCUPANCY_WITH_ALL_EGVS))
```

COMMAND 15
(This command creates a scatter plot of the distance to the edge of the patch of oak woodland vs occupancy and adds a line of best fit.)

```
plot(BLUE_TIT_OCCUPANCY_WITH_ALL_EGVS$DIST_EDGE,
BLUE_TIT_OCCUPANCY_WITH_ALL_EGVS$OCCUPIED, main="OCCUPANCY
  VS DIST_EDGE", xlab="DIST_EDGE",  ylab="OCCUPANCY")
       abline(lm(OCCUPIED~DIST_EDGE, data=
       BLUE_TIT_OCCUPANCY_WITH_ALL_EGVS))
```

Appendix I: R Code Used In The Exercises In This Workbook

COMMAND 16
(This command runs a logistic linear regression with multiple explanatory variables.)

```
GLM<-glm(OCCUPIED ~
ELEVATION+SLOPE+HILLSHADE+DIST_LOCH+DIST_EDGE,
    family=binomial(link='logit'), data=
    BLUE_TIT_OCCUPANCY_WITH_ALL_EGVS)
summary(GLM)
```

COMMAND 17
(This command imports the contents of an attribute table that has been copied to the clipboard of your operating system.)

Before you run this command in R, open the attribute table of the data layer called BLUE_TIT_BREEDING_DATA_WITH_ALL_EVGS in QGIS and copy its contents to your clipboard.

```
BLUE_TIT_BREEDING_DATA_WITH_ALL_EGVS<-read.table(file =
    "clipboard", sep = "\t", header=TRUE)
```

This code creates an object in R called BLUE_TIT_BREEDING_DATA_WITH_ ALL_EGVS which contains the contents of attribute table which you copied from QGIS. This is the name the code below assumes has been used for this data set in R. **NOTE**: If you are using a Mac OS computer, you will need to replace the text `"clipboard"` with the text `pipe("pbpaste")` for this command to work.

COMMAND 18
(This command runs a generalised additive model (GAM) of the relationship between clutch size and multiple explanatory variables.)

```
library(mgcv)
GAM<-gam(NO_EGGS~+s(ELEVATION,fx=F,k=4)
    +s(SLOPE,fx=F,k=4)+s(HILLSHADE,fx=F,k=4)
    +s(DIST_LOCH,fx=F,k=4)+s(DIST_EDGE,fx=F,k=4),
family=poisson, data=BLUE_TIT_BREEDING_DATA_WITH_ALL_EGVS)
summary(GAM)
plot(GAM, pages=1)
```

Appendix I: R Code Used In The Exercises In This Workbook

NOTE: If you encounter any problems when running this code, check that you have the MGCV package downloaded and installed in your version of R. To do this, enter the R code:

```
install.packages("mgcv")
```

Once this package has been installed in R, you can close the R Console and the above command should work as intended in the QGIS R portal.

COMMAND 19

(This command runs a generalised additive model (GAM) of the relationship between the number of hatchlings in each nest box and multiple explanatory variables.)

```
library(mgcv)
GAM<-gam(NO_HATCHED~+s(ELEVATION,fx=F,k=4)
+s(SLOPE,fx=F,k=4)+s(HILLSHADE,fx=F,k=4)
+s(DIST_LOCH,fx=F,k=4)+s(DIST_EDGE,fx=F,k=4),
family=poisson, data=BLUE_TIT_BREEDING_DATA_WITH_ALL_EGVS)
summary(GAM)
plot(GAM, pages=1)
```

COMMAND 20

(This command runs a generalised additive model (GAM) of the relationship between the number of fledglings in each nest box and multiple explanatory variables.)

```
library(mgcv)
GAM<-gam(NO_FLEDGED~+s(ELEVATION,fx=F,k=4)
+s(SLOPE,fx=F,k=4)+s(HILLSHADE,fx=F,k=4)
+s(DIST_LOCH,fx=F,k=4)+s(DIST_EDGE,fx=F,k=4),
family=poisson, data=BLUE_TIT_BREEDING_DATA_WITH_ALL_EGVS)
summary(GAM)
plot(GAM, pages=1)
```

Appendix II: Trouble-shooting Problems With Running Code In R

If you have any problems with running the R code provided in this workbook, these are usually caused by one of the following three possibilities: 1. The data from the attribute table have not been successfully imported into R using the `read.table` command; 2. There is a mis-match between the field names in your data set and the ones provided in the instructions in this workbook (and so the ones that the provided R code is expecting to exist); 3. There is a typo of some kind in the R code you have entered.

You can find out if your data have been imported into R correctly using the `View` command (**NOTE:** Unusually for R, this command starts with an uppercase letter). For example, for the data set imported into R in step 2 of exercise three in this workbook, you can check if the data have been imported correctly by entering the following command into R:

```
View(BLUE_TIT_OCCUPANCY_WITH_ELEVATION)
```

This command will open a window where you can view the data contained in this R object and compare it to the attribute table of the original data set in QGIS. If these do not match, or if you get an error message when you try to use this command, it means your data have not been imported correctly. The best way to deal with this is to repeat the importation step.

To check whether there are any mis-matches between the field names used in your data set and the ones provided in this workbook, you can use the `names` command. For the `BLUE_TIT_OCCUPANCY_WITH_ELEVATION` dataset, this can be done using the following command:

```
names(BLUE_TIT_OCCUPANCY_WITH_ELEVATION)
```

Appendix II: Trouble-shooting Problems With Running Code In

This command will return a list of the field names for your data set, and you can check these against those provided in this workbook. If there is any mis-match, such as the spellings or differences in the use of uppercase and lowercase letters, you will either need to adapt the R code provided, or go back and repeat the appropriate steps in QGIS to ensure that you are using the correct field names (including the correct use of uppercase letters).

To check whether there are any typos within the R code you are using, the only option is to read through the code slowly and carefully, and check each character to ensure that it is correct (including the correct use of uppercase and lowercase letters). If you spot any errors, these need to be corrected before you try to run the R code again, and it can often take several attempts before you have successfully found and eliminated all such errors.

Appendix III:
How To Transfer Data From R To QGIS

In this workbook, you have learned how to transfer data from QGIS to R in order to conduct spatial analyses. However, there may also be occasions where you wish to transfer data from R and to QGIS, such as a table showing the results of a particular data analysis. This can be done using the same single storage file system and common data format outlined on page 4. The exact data format which you use will depend on your own specific requirements, but in most instances, the easiest way to do this is to export your data from R using the .CSV (comma separated values) format. This is because this file format can easily be added to QGIS using the ADD DELIMITED TEXT LAYER tool (this can be found by clicking on LAYER on the main menu bar and selecting ADD LAYER> ADD DELIMITED TEXT LAYER). To export a table object from R as a .CSV file, you can use the `write.table` tool.

For example, in step 4 of exercise three, you created a new R object called NESTBOX_OCCUPANCY. This was a table of the occupancy rates of nest boxes by blue tits in five elevation categories. If you wished to export this object from R so that it could imported into QGIS, you would enter the following command into your R CONSOLE window:

```
write.table(NESTBOX_OCCUPANCY,
file="C:/QGIS_R_WORKBOOK/NESTBOX_OCCUPANCY.CSV", sep=",",
dec=".")
```

NOTE: In order to be able to plot a data set exported from R in QGIS, it will need to have fields that contain spatial coordinate values, such as latitude and longitude.

Lightning Source UK Ltd.
Milton Keynes UK
UKHW030619240220
359221UK00004B/30